THE SHORTEST HISTORY OF THE DINOSAURS

ALSO BY RILEY BLACK

When the Earth Was Green

The Last Days of the Dinosaurs

Skeleton Keys

My Beloved Brontosaurus

Written in Stone

THE
SHORTEST
HISTORY
OF THE
DINOSAURS

**The 230-Million-Year Story
of Their Time on Earth**

RILEY BLACK

NEW YORK

THE SHORTEST HISTORY OF THE DINOSAURS: *The 230-Million-Year Story of Their Time on Earth*
Copyright © 2025 by Riley Black
Pages 215–18 are a continuation of this copyright page.

All rights reserved. Except for brief passages quoted in newspaper, magazine, radio, television, or online reviews, no portion of this book may be reproduced, distributed, or transmitted in any form or by any means, electronic or mechanical, including photocopying, recording, or information storage or retrieval system, without the prior written permission of the publisher.

The Experiment, LLC
220 East 23rd Street, Suite 600
New York, NY 10010-4658
theexperimentpublishing.com

THE EXPERIMENT and its colophon are registered trademarks of The Experiment, LLC. Many of the designations used by manufacturers and sellers to distinguish their products are claimed as trademarks. Where those designations appear in this book and The Experiment was aware of a trademark claim, the designations have been capitalized.

The Experiment's books are available at special discounts when purchased in bulk for premiums and sales promotions as well as for fundraising or educational use. For details, contact us at info@theexperimentpublishing.com.

Library of Congress Cataloging-in-Publication Data available upon request

ISBN 979-8-89303-056-3
Ebook ISBN 979-8-89303-057-0

Cover and text design by Beth Bugler

Manufactured in the United States of America

First printing June 2025
10 9 8 7 6 5 4 3 2

For Joey
My little sunspot

Contents

Key Moments in the History of the Dinosaurs	viii
Major Dinosaur Discoveries on Every Continent	x
Introduction	1
1. HOW TO MAKE A TERRIBLE LIZARD	19
2. EGGS AND NESTS	39
3. GROWING UP DINOSAUR	53
4. HOT-RUNNING DINOSAURS	69
5. THE LARGEST CREATURES TO WALK THE EARTH	85
6. DINOSAURS OF A FEATHER	97
7. DINOSAUR DIETS	113
8. HOW DINOSAURS MADE THEIR WORLD	125
9. DECORATION AND DEFENSE	137
10. THE SOCIAL DINOSAUR	151
11. DINOSAURS UNDONE	165
12. HOW TO BECOME A FOSSIL	181
Afterword	197
Further Reading	203
Image Credits	215
Acknowledgments	219
Index	221
About the Author	227

Key Moments in the History of the Dinosaurs

TRIASSIC
251–201 million years ago

251 MYA – The end-Permian mass extinction, caused by volcanic eruptions: 81% of marine species and 70% of terrestrial vertebrates go extinct.

232 MYA – The earliest dinosaurs, like *Nyasasaurus*, roam ancient Africa and South America in southern Pangaea.

229 MYA – The earliest members of the major dinosaur groups Saurischia and Ornithischia evolve.

201 MYA – The end-Triassic mass extinction, caused by volcanic eruptions: Crocodile relatives are devastated but dinosaurs are virtually unscathed.

JURASSIC
201–143 million years ago

200 MYA – The supercontinent Pangaea begins to break apart, into northern and southern parts.

199 MYA – The earliest sauropods evolve, like *Antetonitrus* from prehistoric Africa.

180 MYA – The southern supercontinent Gondwana begins to break apart, as prehistoric Africa and South America split.

166 MYA – The oldest ankylosaurs evolve, such as *Spicomellus* from prehistoric Africa.

165 MYA – *Megalosaurus*, the first dinosaur to be named by scientists, roams ancient England.

161 MYA – The earliest horned dinosaurs, such as *Yinlong*, evolve in the Northern Hemisphere.

150 MYA – Droughts and local floods create vast bone beds, such as those in Dinosaur National Monument in Utah and Mother's Day Quarry in Montana.

149 MYA – Feathered dinosaurs give rise to the first bird, *Archaeopteryx*.

CRETACEOUS
143–66 million years ago

125 MYA – The oldest angiosperms, or "flowering plants," evolve in eastern Asia.

123 MYA – Fuzzy dinosaur *Sinosauropteryx* and many other creatures are preserved in northeastern China among what's known as the Jehol Biota.

110 MYA – Hadrosaurs evolve, such as the early duckbill *Equijubus*.

100 MYA – The Western Interior Seaway begins to split North America in two by washing over the middle of the continent, and lasts until the end of the Cretaceous.

80 MYA – The earliest tyrannosaurids evolve, large predators including the ancestors of *Tyrannosaurus*.

75 MYA – Coastal habitats from Alaska to Mexico preserve dinosaur faunas including tyrannosaurs, horned dinosaurs, ankylosaurs, hadrosaurs, and more on the North American subcontinent Laramidia.

66 MYA – The end-Cretaceous extinction, caused by asteroid impact: More than 75% of species perish, including all dinosaurs except birds.

Major Dinosaur Discoveries on Every Continent

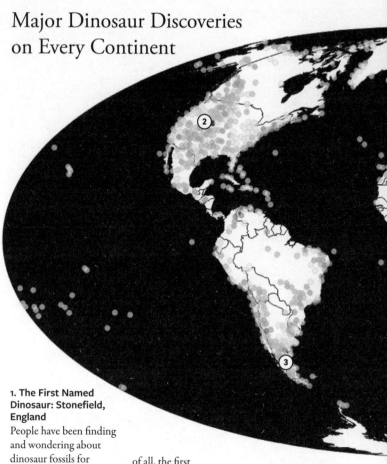

1. The First Named Dinosaur: Stonefield, England

People have been finding and wondering about dinosaur fossils for centuries, but the first dinosaur to receive a scientific name was the carnivore *Megalosaurus*. It was named in 1824 from fossils found in England.

2. The Tyrant King: Eastern Wyoming, USA

Arguably the most famous dinosaur of all, the first recognized skeleton of *Tyrannosaurus rex* was found in eastern Wyoming in 1900. The skeleton was one of more than fifty partial skeletons now known.

3. Giant Among Giants: Chubut Province, Argentina

The long-necked dinosaur *Patagotitan* is the current contender for largest dinosaur of all time. In life, the dinosaur may have measured over 100 feet long and weighed over 60 tons.

4. Oldest Dinosaur: Ruhuhu Basin, Tanzania

The oldest dinosaurs yet known have been found

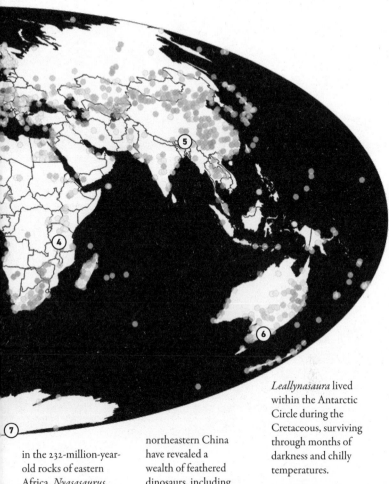

in the 232-million-year-old rocks of eastern Africa. *Nyasasaurus*, a beetle-eater about the size of a German shepherd, is the current contender for earliest dinosaur.

5. Prehistoric Aviary: Laioning, China
Fossil beds in northeastern China have revealed a wealth of feathered dinosaurs, including *Sinosauropteryx*. In 1993, the fuzzy dinosaur was the first non-avian dinosaur found with intact plumage.

6. Dinosaurs in the Dark: Victoria, Australia
The small dinosaur *Leallynasaura* lived within the Antarctic Circle during the Cretaceous, surviving through months of darkness and chilly temperatures.

7. Crested Carnivore: Transantarctic Mountains, Antarctica
The crested meat-eater *Cryolophosaurus* was the first dinosaur named from Antarctica, its name meaning "frozen crest lizard."

Introduction

We still live in the Age of Dinosaurs. The sentiment might seem strange given that we most often associate the word *dinosaur*—literally "terrible lizard"—with primordial forests where scaly, long-fanged carnivores stalk and many-horned giants waddle through fern-covered groves in an ancient, endless summer. But the dinosaurs are still here, and not only as mineralized bones spilling out of stone or reassembled in museum halls. All around us, perhaps just outside your window or in the nearest tree, one of the "terrible lizards" perches. As the last surviving dinosaur lineage, birds are continuing an ancient legacy that began in the wake of Earth's worst mass extinction. A chickadee or pigeon is a thread of an evolutionary tapestry that stretches back over 235 million years.

Our understanding that birds are all that's left of the dinosaurs is a relatively recent revelation. Paleontologists now mark the division between avian dinosaurs, or birds, and non-avian dinosaurs—animals such as the famous *Stegosaurus*, *Diplodocus*, and, of course, *Tyrannosaurus*. The split is a scientific symptom of how old our fascination with dinosaurs is, that we began to study and categorize them before we had any idea they were connected to our everyday lives. In fact, we can trace the glimmerings of our persistent affection for dinosaurs back beyond the date when the reptiles were scientifically named. Our fixation did not begin with the moment the word "dinosaur" was invented by an English anatomist in 1842. People had been uncovering and wondering about dinosaurs for thousands of years by that point. In South Korea, Brazil, and the United States, ancient rock artworks called petroglyphs are sometimes found in close association with prehistoric dinosaur footprints. At one site, in Utah's Zion National Park, some of the petroglyphs are even shaped like the characteristic, three-toed footprints left by carnivorous dinosaurs around two hundred million years ago. Indigenous peoples the world over recognized that dinosaur footprints, as well as bones and other fossils, were left by unknown creatures from the distant past, bringing them into their cultures long before paleontology even became a science. The fossil footsteps, teeth, and bones non-avian dinosaurs left behind were potent clues that our world was once home to creatures far different from anything humans had seen alive. Imagine coming across the exposed skeleton of a dinosaur like the three-horned *Triceratops* if you had no understanding of what a *Triceratops* or a dinosaur was. What would you make of the giant bones?

Paleontology did not arise from the recognition that dinosaur bones and footprints represent once-living animals, however. The science intellectually sprouted and was formalized in Western Europe through the eighteenth and nineteenth centuries, requiring that naturalists and experts dismantle some philosophical baggage around the age of the Earth and the true nature of what fossils are. Fossils, prior to the 1700s, were not generally recognized as having anything to do with prehistoric life at all. Influenced by Christianity, the prevailing intellectual climate insisted that the world was relatively young and that every living thing had existed from that time in a state of balance. Nothing new evolved, nothing went extinct, and the world simply wasn't old enough for there to be mineralized remains of ancient creatures. People certainly found fossils, but the fact that fossils seemed to be stone animal parts had to be explained away in order for Earth's youth and stability to be maintained. Fossil shark teeth commonly found in rocks laid down by ancient seas, for example, were not seen for what they clearly were, but had been deemed to be curiosities that fell from the moon by the Roman naturalist Pliny the Elder in 77 CE. The same triangular fossils were later colloquially called *glossopetrae* and said to be the petrified tongues of snakes Saint Paul had turned to stone. Even by the 1600s, as the science of geology began to take shape, German scholar Athanasius Kircher still attributed the fossils to vital forces within the Earth that mimicked life. Fossils were not remains of living things, he proposed, but were natural fakes that were just one form of geological wonder among many, such as stones that superficially resembled eyes or eggs. Any explanation seemed more acceptable than the earthshaking idea that the world was far older than what the chronologies of the Biblical patriarchs suggested, leading scholars to do mental backflips to explain why what clearly looked like shark teeth must be attributable to some other source. It wasn't until 1616 that Italian naturalist Fabio Colonna demonstrated that the shark teeth were organic in origin, not stone, and not until 1669 that Danish scientist Niels Steensen worked out the geological basics of how such shark teeth could end up preserved high and dry in Europe's mountains. The teeth must have been shed into sediment that was covered by additional layers and later exposed, he proposed, which would also require that the Earth be old enough for such vast changes to unfold.

Despite the fact that people around the world had recognized fossils as remnants of ancient life over thousands of years, it wasn't until the seventeenth century that European naturalists were prepared to accept the reality of fossils. Even then, it was still another century before European experts would acknowledge that some of these fossils represent species that have vanished—that extinction was a reality, despite the fact that colonizers had already been responsible for the disappearance of species like the dodo. The very idea that species might entirely disappear, or might have been different in the distant past, seemed outlandish. But fossils don't lie. Enslaved people brought to North America recognized the teeth and bones of mammoths and mastodons as similar to those of elephants in Africa. It took naturalists years of study to come to the same conclusion. In 1799, the French anatomist Georges Cuvier proposed that huge bones from the United States called the "American incognitum" and equally impressive remains from Siberia represented two entirely extinct elephants, the mastodon and mammoth, that could be reliably told apart by their anatomy. The colonial push that was at the time heralded as an age of exploration left little doubt that someone surely would have seen a living mastodon or mammoth if they still lived, and so Cuvier provided proof of what only a few naturalists had dared speculate about before. Nature never existed in a state of perpetual and self-sustaining balance, but had seen the arrival and departure of many different species, each of which now raised questions about why such bizarre life forms appeared and why they ultimately became extinct.

People were still finding dinosaurs during this time. Naturalists and collectors excavated and illustrated dinosaur bones, often mislabeling them as parts of big crocodiles that spread across the world when it was warmer or as fragments left behind by Roman war elephants. Naturalists had to recognize that the Earth is older than presumed, that species do go extinct, and that Earth's array of species changed over time before they could even begin to comprehend the existence of something like a dinosaur. The rock record of western Europe did not do them any favors. If people had invented the science of paleontology somewhere else on the planet, such as China or Argentina, where relatively complete dinosaur fossils have a better chance to be found, perhaps dinosaurs and their role in Earth's history would have been recognized sooner. As it was, the dinosaur-preserving rocks of western Europe most often yielded isolated bones, teeth, and fragments. No one would find a complete skeleton that depicted the full form of a dinosaur until 1850. Without such a fossil, paleontologists of the early nineteenth century did not even realize that creatures we now know as dinosaurs even existed.

The emergence of the dinosaurs was slow and had to begin with the familiar. Naturalists, anatomists, and the intellectual forerunners of paleontologists started with shark teeth like those still seen in living fish, prehistoric elephants similar to those still living in Africa and Asia—and worked backward. Fossils had to be associated with still-living organisms to be recognized, and those that didn't seem to correspond to known animals were often treated as mysteries or even parts of myth. Jurassic dinosaur footprints found by colonizers of North America's northeast in the early nineteenth centuries were sometimes attributed to "Noah's raven," or birds that survived the Biblical Deluge, and were cataloged as "sandstone birds" by Massachusetts naturalist Edward Hitchcock in the 1830s. Only birds were known to make such distinctive,

INTRODUCTION 5

three-toed tracks, and there were no skeletal remains in the sandstone layers to indicate what creatures made the footprints. "Bird" was a best guess and there was no reason to think there was some other form of animal that would leave such similar traces behind.

Nevertheless, the simple recognition that Earth and life on it were far older than anyone had dreamed sent paleontologists looking for more clues. Animals from the not-so-distant past, such as the mastodon and the giant ground sloth *Megatherium*, were clearly associated with living animals despite being different from any modern species. The eventual discovery of the giant seagoing lizard *Mosasaurus* suggested that life diverged even further the older it got. A jumble of jaw and skull bones were found in a Netherlands chalk mine in 1764 and, despite the emergence of a second, better preserved skull found in the same mine, the bones were thought to be those of either a big crocodile or a whale. It wasn't until 1800 that the bones were recognized as those of an enormous reptile related to monitor lizards, a fully aquatic form that lived even before the mastodon and *Megatherium*. The animal was eventually named *Mosasaurus*, and even stranger marine reptiles were found in the Jurassic rocks of southern England around the same time. The long-necked, four-paddled *Plesiosaurus* and the sharklike *Ichthyosaurus*, both named in 1821, were better-suited to living in the seas than any known reptile, as they were fundamentally different from living lizards and crocodiles. The finds left no doubt that life grew ever stranger the deeper into the fossil record paleontologists looked. Finally, with the recognition that unusual and giant reptiles had once lived on Earth in place, dinosaurs were ready to make their scientific debut.

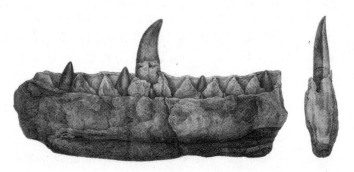

The lower jaw of *Megalosaurus* studied by William Buckland, who named the reptile.

Even though dinosaur fossils were already being found in other parts of the world, the fossils critical to the origin of dinosaurs as an *idea* were found in England. Quarry workers, naturalists, and other curious people had been finding giant teeth and bones in the English countryside for years. No one was entirely sure what they came from. Perhaps they belonged to enormous crocodiles that lived in the area long ago. Theologian William Buckland was especially intrigued by these remains, and, in correspondence with Cuvier in France, realized that some must have belonged to huge reptiles. A segment of fossil jaw with a serrated, knife-like tooth, in particular, seemed to represent a carnivorous saurian unlike any living, which Buckland named *Megalosaurus*—"giant lizard"—in 1824. We know it today as the first dinosaur to receive a scientific name. More enigmatic reptiles soon followed. During the same time period that Buckland was pondering his fossil collection, and during a stroll in the English countryside with her friend Bridget Waller, Mary Ann Mantell spotted some especially interesting fossils in the freshly broken stone of a quarry and bought them from the workers there. Mary Ann's husband, Gideon, was a surgeon with a growing interest in fossils, and so she brought them back for him to study. The fossils were teeth, but from what sort of creature? Gideon Mantell carefully pursued the question over the next several years, asking other experts and carrying out his own comparisons. The teeth looked very much like those of an iguana, but from a giant animal that likely left some of the enormous bones found around the countryside that Mantell was already aware of. Mantell eventually named the small collection of fossil bones *Iguanodon*.

Megalosaurus and the plant-munching *Iguanodon* were not envisioned as the dinosaurs we know them as today. Buckland, Mantell, Cuvier, and other experts of the time saw these creatures as huge equivalents of modern reptiles. *Megalosaurus* was probably a large and unusual crocodile, they proposed, and *Iguanodon* likely looked like a green iguana—albeit one 100 feet long (30 m). But as British anatomist Richard Owen studied these fossils, as well as those of an armored reptile that Mantell named *Hylaeosaurus* in 1833, he noticed that each shared some key traits in common to the exclusion of other saurians. The limb bones of the three creatures all indicated that they held themselves up on upright, column-like limbs rather than with legs sprawled out to the sides like lizards do. *Megalosaurus*, *Iguanodon*, and *Hylaeosaurus* all had several fused vertebrae between their hips, as well, in an arrangement not seen before. No other reptiles, living or fossil, were known to have skeletal features like this, and so Owen grouped them together. *Megalosaurus*, *Iguanodon*, and *Hylaeosaurus*, he pronounced in an 1842 monograph of a talk he'd given the year before, belonged to the same group of extinct reptiles—the Dinosauria.

The enormity of what Owen had just found was unknown even to him. No one had any concept of just how many dinosaurs there were, or what the reptiles truly looked like in life. The dinosaurian connection to birds wasn't even a thought in anyone's mind yet as there seemed to be no resemblance between a hulking *Iguanodon* and the seagulls flying over England's Jurassic Coast. The three dinosaurs Owen had identified were represented by only bits and pieces, some of which have now been categorized as entirely different species. Owen saw each of his fossil trio as elevated reptiles, almost as if reptiles were becoming more mammal-like. (While he rejected evolution by natural selection, Owen held evolutionary ideas of his own based around increasing perfection from a primordial archetype.) He even got some assistance bringing this image of the dinosaurs to the public, working with artist Benjamin Waterhouse Hawkins to create life-size versions of rhino-like dinosaurs that still stand at London's Sydenham Hill today. Despite these errors and misinterpretations, Owen's assessment that dinosaurs were *different* from other reptiles was solid. Naturalists and academics began finding creatures with the same basic traits in other parts of the world, especially as students trained in Europe began to establish centers of fossil study in North America. One of the most pivotal finds was made in 1838, before anyone could have realized its full significance. Farmer John Estaugh Hopkins was digging in the sandy soil of his Haddonfield, New Jersey farm when he found the bones of some huge creature. He kept the bones on display for two decades before amateur geologist William Parker Foulke saw them, went back to the discovery site, and uncovered even more, informing Philadelphia-based naturalist Joseph Leidy about the find. In 1858, Leidy described the bones as

Hadrosaurus—an herbivorous dinosaur that had longer hind limbs than forelimbs and walked in a way no living reptile did.

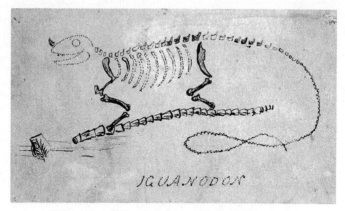

Gideon Mantell sketched the possible form of *Iguanodon* based on the assumption that the extinct reptile had proportions similar to living iguana lizards.

Even though Europe's new geologists and paleontologists knew there were fossils elsewhere around the world, often brought to their doorstep by the extractive methods exercised through colonialist occupation, experts generally expected fossils found abroad to be like those in Europe. No one knew the absolute age of the Earth, geologists would not accept the reality of continental drift for almost another century, and somehow naturalists thought they were seeing a "progression" of life through the ages, in which bizarre reptiles like dinosaurs were temporary placeholders that came after the emergence of the amphibians but before mammals. The reptiles were too few and too strange for scientists to understand their connection to other forms of life, leaving them as a strange interlude in broader narratives of change. Science had introduced the world to the dinosaurs, but no one really knew how they fit into the Tree of Life, or even what they really looked like. As what had become the United States continued its campaign of genocide and occupation of western North America, however, railways soon gave government surveys and scientists access to broad expanses of fossil-packed rock layers. The early finds were fragmentary, perhaps because no one yet knew what to look for, but soon the deserts of western North America would provide a vast array of new dinosaurs that not only changed the image of what the reptiles were like, but sparked a tense scientific competition to find the biggest, best, and strangest of all.

As the United States pushed its boundaries to the west, railroads snaking across the country, colleges and museums in the East gave academic homes to a new generation of paleontologists and geologists ready to receive fossils dug up on military expeditions and by settlers. The story was bigger than that of the infamous "Bone Wars" between Philadelphia naturalist Edward Drinker Cope and Yale University's Othniel Charles Marsh, independently affluent researchers who simultaneously set up the next generation of American paleontologists while publicly quarreling to the point of embarrassing the entire field. The competition between the two certainly did yield exquisite skeletons of creatures such as stubby-armed *Ceratosaurus*, armored *Stegosaurus*, long-necked *Apatosaurus*, and more, not to mention various fish and mammals and other prehistoric creatures, but the legendary fight between the two often overshadows the close ties between early paleontology and conquest. Just as Richard Owen received fossils from South Africa and India due to England's occupation, so did Cope, Marsh, Leidy, and others benefit from the greedy legacy of "Manifest Destiny" as the boundaries of the United

States stretched from coast to coast. The deep geological history of the American West had entombed many forms of ancient life and shifted to bring those fossils close to the surface in broad geological swaths. Army expeditions and miners searching for geological riches like silver often found fossil spots first. Now paleontologists and the public alike were being introduced to creatures such as *Triceratops*, the iconic three-horned herbivore whose remains were originally mistaken for those of a bison, and tiny-headed, long-necked giants such as *Diplodocus*, visualized in some newspapers as standing up against a skyscraper and peering into the upper windows. Not only were dinosaurs weird, but they came in a wider array of forms than anyone expected.

The race to find America's dinosaurs that began in the late nineteenth century spilled over into the early twentieth as new museums were constructed as cathedrals of science and knowledge. Experts know this as the Second Jurassic Dinosaur Rush. Institutions such as New York City's American Museum of Natural History, Pittsburgh's Carnegie Museum of Natural History, and Chicago's Field Museum all sought their own showstopper dinosaur, sending their experts to the Jurassic rocks of Wyoming, Colorado, and Utah to bring back something big and impressive enough to draw in visitors. All of them were successful, to varying extents, often finding new species along the way. The Carnegie's efforts, in particular, turned up a skeleton of *Diplodocus* so exquisite that not only was it recognizable as a new species, *Diplodocus carnegii*, but Carnegie himself had multiple plaster copies of the skeleton made so he could gift them to the leaders of other countries as he traveled abroad. Carnegie had effectively bought a dinosaur, down to its very name. Dinosaurs were not merely objects of scientific interest, but were symbols of colonialist acquisition, industrialist ideals, and power.

Colonialism fueled paleontology, as the search for minerals and natural resources revealed bones that were funneled back to experts in Western Europe and the United States. In the early 20th century, German engineer Bernhard Wilhelm Sattler exploited local knowledge and labor in Tanzania to send large Jurassic dinosaur bones back to Berlin.

No matter how many dinosaurs experts unearthed, however, the reptiles stood apart from other forms of life. It wouldn't be a stretch to say that many paleontologists were bored with dinosaurs, finding them useful to impress visitors and journalists but otherwise of relatively little scientific interest. Fossil mammals, by contrast, were abundant and had ties to the beasts of the modern world, allowing paleontologists to think about the mechanics of evolution. After all, Darwin's proposal of evolution by natural selection was widely derided for decades, and throughout the late nineteenth and early twentieth centuries, paleontologists drew from fossil mammals and other creatures to come up with alternative ideas, especially involving vital forces that seemed to nudge living things toward advanced and improved forms over time. The still-growing menagerie of dinosaurs was thought to be irrelevant to evolutionary considerations, representing only an odd interlude before mammals took over. Even by the 1940s, when scientists in several biological disciplines realized that natural selection really is the main engine of evolution, dinosaurs played no part in the considerations. They were big reptiles and probably acted like big lizards and crocodiles, the logic went, despite the fact that biologists of the twentieth century knew vanishingly little about living reptiles, too.

But throughout it all, some experts remained curious about dinosaurs. Small, carnivorous species, such as the Jurassic "bird robber" *Ornitholestes*, couldn't have possibly been sluggish, dim-witted creatures. Everything about their anatomy suggested that such dinosaurs were adept hunters, and perhaps warm-blooded. Some paleontologists, too, saw a close resemblance between the skeletons of toothed early bird *Archaeopteryx*, first uncovered in 1859, and small carnivorous dinosaurs, hinting that birds evolved from something dinosaur-like if not the dinosaurs themselves. Experts had found too many dinosaurs of all shapes and sizes for them all to be swamp-bound slowpokes, and by the 1960s, a new generation of paleontologists began questioning the standard story. Fossil studies in general were going through a "paleobiological revolution" in which experts began asking questions about body temperature, social habits, reproduction, and evolution, the increasing accessibility of computers allowing paleontologists to do everything from recognize the reality of mass extinctions to compare the shapes of individual fossils and uncover whether they might be different sexes or species. The shift was driven by debate and discovery alike. As experts argued over whether some dinosaurs were warm-blooded, finds at locales in Montana and Alberta provided new evidence that some were social and cared for their young. And while the idea of dinosaurs that kept their body temperatures hot had been entertained before, the 1969 announcement of the "terrible claw" *Deinonychus*—a small carnivore with extendable claws on the second toe of each foot, a counterbalancing tail, and birdlike traits all over the skeleton—helped catalyze discussion over whether paleontologists had been misunderstanding the bones all along. It was time for the Dinosaur Renaissance, a burst of scientific fascination that fundamentally transformed dinosaurs and continues to this day. Dinosaurs went from a small and generally uninspiring area of study during much of the twentieth century to a topic that fills multiple sessions over several days at academic conferences such as the Society of Vertebrate Paleontology, not to mention the flood of new research papers that are nearly impossible to keep up with. On average, a new dinosaur species is named every two weeks. Paleontologists may keep that pace for some time. Based upon the number of known dinosaurs and the expanses of possibly dinosaur-bearing rocks that have yet to be explored, most dinosaur species preserved in Earth's sediment have yet to be discovered. If that sounds strange, consider how long dino-saurs have existed on our planet. A little more than sixty-six million years separates us from the last *Tyrannosaurus*. About 150 million years, however, have passed since the plate-backed *Stegosaurus* walked over Jurassic floodplains. The two famed dinosaurs lived more than eighty million years apart, in which we could fit the entirety of the planet's history since the time

INTRODUCTION   15

Ongoing discoveries have fundamentally changed what paleontologists think dinosaurs were like, especially early finds such as *Iguanodon*.

of *Tyrannosaurus*, with plenty of room to spare. And that's not even reaching back to the earliest dinosaurs that strutted around the Earth more than 230 million years ago. To put it another way, bones of dinosaurs like *Stegosaurus* had already been fossils for tens of millions of years, possibly even being exposed and eroded away tens of millions of years before the earliest humans even existed. The span of time dinosaurs have existed on Earth is literally incomprehensible.

What began as curiosity about confusing, almost mythical creatures has transformed into a thriving subdiscipline of paleontology all its own, dinosaurs standing out as the most iconic and beloved out of all the wonderful forms of life our planet has ever hosted. Dinosaurs are practically synonymous with prehistory itself despite representing only a tiny fragment of life's biodiversity. By looking at a *Velociraptor* in a book or an *Apatosaurus* in a museum hall, we're immediately confronted with the fact that the Earth is ancient, living things dramatically changed through that time to the present, and the vast majority of creatures that have ever existed are now extinct, the mysteries always outnumbering what we know. Perhaps this is why we pursue dinosaurs. For every new fact we learn about them, we come up with a litany of new questions we deeply want answers to.

What follows throughout the rest of this book is the story, so far as we presently understand it, of how dinosaurs came to be such influential aspects of Earth's ever-changing ecosystems. Experts have been able to assemble an understanding of when and why the first dinosaurs evolved, what led them to become so numerous and widespread, and why nothing like *Triceratops* walks among us today. Dinosaur social habits, courtship displays, sounds, and even colors have come into focus, as well, enabling us to envision dinosaurs in the flesh more precisely than ever before. New research will continue to adjust and correct the tale, but we now know enough about some essential prehistoric moments to understand how dinosaurian history unfolded.

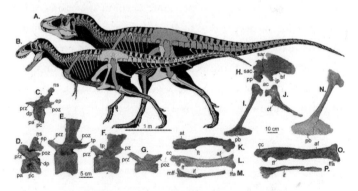

Paleontologists are continuing to uncover new dinosaur species at an incredible rate, such as new tyrannosaur species from parts of the American Southwest that were previously overlooked.

Summarizing more than 235 million years of evolutionary history, not to mention more than a half century of research, is a challenging task. Dinosaurs are best understood in the context of our ever-changing world, involving the slow grind of tectonic plates, constant shifts in climate, the evolutionary histories of everything from plants to parasites, and even, as we well know, the nature of our planet in a solar system where great chunks of rocky debris sometimes zip over for a catastrophic visit. We are still very much in the process of uncovering the full history of the dinosaurs, and so by the very nature of the subject matter, this book can be only an incomplete look at the evolutionary epic that dinosaurs represent. Nevertheless, it is truly incredible how much we have come to understand about creatures that thrived and perished more than sixty-six million years before our own time. Animals that once existed only in the realm of myth and unknowable times are now connected to the pigeon on the street and the hummingbird sipping from a flower, not a strange chapter in life's story but integral to the history of the world as we know it. Our world wouldn't be the same without dinosaurs, and our own fate is deeply tied to that of these amazing reptiles. Where we can follow ancient footprints and touch prehistoric bone, we are making connections to an epic that is still unfolding. Dinosaurs not only shaped a world our ancestors lived in but also remain with us to this day.

CHAPTER 1

How to Make a Terrible Lizard

Dinosaurs appeared in the wake of disaster. Paleontologists know it as the end-Permian mass extinction that peaked 252 million years ago. The catastrophe, often called the worst mass extinction of all time, was necessary for there to be an "Age of Reptiles" at all. For tens of millions of years before the mass extinction, protomammals were the most numerous and diverse vertebrates on land. Reptiles evolved alongside them but did not reach the same array of shapes and sizes. The mass extinction, which wiped out more than 75 percent of known species on land, nearly eradicated the protomammals entirely and offered the surviving reptiles the opportunity to undergo their own evolutionary fallout. Without this terrible event, dinosaurs may never have evolved at all.

The consequences of this world-changing phenomenon can be understood only through comparing what life was like on either side of the mass extinction. Prior to 252 million years ago, there was not even a whisper of a possibility that enormous, armored, and feathery reptiles would one day be some of the most numerous and diverse creatures on the planet. Earth's land-based ecosystems hosted an extremely different menagerie of creatures. Reptiles were part of these ancient assemblages, from snaggletoothed carnivores to hulking plant-munchers, but their variety was far surpassed by our relatives and ancestors—the synapsids, or protomammals.

Throughout evolutionary time, synapsids have been united by a particular window in the skull. Synapsids have an opening on the side of the skull, behind the eye, called the temporal fenestra. (Humans even retain this feature in the form of the opening made by the arch of our cheekbones.) Most reptiles, by contrast, are called diapsids and are distinguished by two holes behind the eye. Given that synapsids comprise mammals and all their prehistoric relatives more closely related to them than reptiles, the early synapsids of the Permian period are often called protomammals.

The first synapsids were lizard-like creatures. Many of them didn't look like mammals at all. The iconic, sail-backed *Dimetrodon* was a synapsid, for example, and more closely related to us than to any reptile. And throughout the Permian, approximately forty-seven million years, synapsids changed from roughly lizard-like animals to a strange array of fuzzy creatures that seemed to combine both reptilian and mammalian traits. The creatures evolved and spread across the supercontinent Pangaea, a vast expanse that included the majority of the world's landmasses—the stitched-together parts of the Earth's crust that would later become the continents we reside on today. In a relatively cool global climate, the interior of Pangaea was comparatively arid and rain more often fell near the coast. Seasons in the Permian world were sharply distinct from one another, too, with parched dry seasons broken by the arrival of megamonsoons that marked the wetter months. Ferns and early conifer relatives dotted the landscape, along with new arrivals like cycads and ginkgoes. It was in this world that an array of synapsids thrived. Some of these protomammals, like the dicynodonts, were herbivorous creatures, built like pigs, with beaks and tusks sticking out of their faces. The biggest carnivores of the time were not reptiles, but animals like *Inostrancevia*—a predator roughly the size of a Great Dane with long saber teeth, roaming across great expanses of Pangaea. Our ancestors were there, too. Weasel-like cynodonts burrowed into the ground, ate plants and bugs, and included some forms that began to look quite mammal-like, gaining more upright postures and greater spine flexibility to move faster and more efficiently than their squat, sprawling forerunners.

If major ecological disruptions had been avoided, the synapsids likely would have remained widespread, diverse, and disparate, part of an alternative evolutionary timeline that would have unfolded as Pangaea gradually split apart. By the end of the Permian, life on Earth had interacted and evolved for over one hundred million years without a mass extinction to reshuffle the deck. On land and in the sea, the Permian world boasted ecosystems brimming with many different species—from trilobites crawling over corals and sponges in the shallows to bear-size predators like *Inostrancevia* sinking their teeth into dicynodonts on land. Life had created complex, ever-changing ecologies over an immense span of time, but much of that biodiversity began unraveling as Earth's geology made the planet near-inhospitable to animals, plants, and many other forms of life.

The end-Permian mass extinction was a long, grinding disaster, not nearly so quick and direct as the asteroid strike that would later culminate the heyday of the dinosaurs at the end of the Cretaceous period. The catastrophe's trigger came from within the Earth. In what's now Siberia, volcanoes oozed out an incredible amount of molten rock, spewing greenhouse gases such as carbon dioxide and methane into the air. The level of carbon dioxide, for example, jumped from about 400 parts per million (ppm)—a little less than today—to 2,500 ppm, an amount that made the air harder to breathe, caused ocean waters to become more acidic, and drove rapid global warming. The rapid rise in temperature caused other reservoirs of greenhouse gases, such as frozen methane in the ocean's depths, to be released and further intensify the ecological upheaval. Even more directly, the prolonged volcanic activity may have even ignited coal seams created by vast, plant-packed swamps that thrived in earlier eras, the fossil record itself fueling the change.

Paleontologists and geologists are still investigating and debating how the resulting mass extinction played out. Some experts see one main pulse, with others pointing to evidence that there may have been as many as three over the course of several million years. Volcanic eruptions in prehistoric China around 259 million years ago, as well as falling sea levels as the climate cooled, are thought by some experts to have caused a significant extinction pulse that destabilized Earth's ecosystems and made them more vulnerable to the more intense eruptions at the end of the Permian. Others think this earlier extinction phase was more of a local disturbance than a global one. Nevertheless, whether earlier events played into the end-Permian mass extinction or there was just one intense event, Earth's biodiversity was hit hard by the persistent shifts to the climate, atmosphere, and seas. Experts estimate that about 81 percent of marine species vanished by the end of the Permian, as well as about 70 percent of vertebrates living on land. The last of the trilobites disappeared, insects endured the worst mass extinction they've ever faced, and, among other losses, the great array of protomammals across Pangaea was reduced to just a few members of two groups—the dicynodonts and the cynodonts.

Life took millions of years to bounce back after the disaster. In the earliest days of the following Triassic period, ecosystems the world over were inhabited by a relatively small number of survivor species that hung on. Forests that had boasted many different species of plants, insects, animals, and more were now monotonous groves of tree ferns that only a few animals seemed to live in. On top of the losses, Earth's climate had grown incredibly hot compared to how it had been before the eruptions, with harsh conditions that pushed organisms into new environments and ways of living. Many reptiles that had survived the mass extinction, for example, found some refuge by the water, adapting to oceans that now lacked any large predators to feast on them as naïve swimmers. Reptiles became some of the first vertebrates to reenter the seas since their ancestors crawled out of it, opening up new evolutionary avenues. Reptiles on land were beginning to undergo an evolutionary consequence of their own, as well. The way reptiles reproduced set them up for an adaptive radiation in a devastated world. Not only did reptiles lay clutches of multiple eggs at a time, but they grew fast and reached sexual maturity before they grew to their full body size. Reptiles were able to have more offspring and faster than the surviving synapsids. The reptiles of the earliest Triassic were able to effectively flood Earth's habitats with their offspring while the surviving protomammals reproduced more slowly, and so reptiles came to be especially prominent in this new Triassic world. The evolutionary stage was finally set for the dinosaurs.

Paleontologists will probably never find the very first dinosaur to ever roam our planet. Evolution takes place in populations, after all, and the differences between the first dinosaur and its non-dinosaur parent were probably so small as to be undetectable. The somewhat sparse nature of the fossil record actually helps paleontologists by creating gaps so that different forms can be more easily compared and associated with each other. Still, it's difficult to look at the imposing skeleton of a *Tyrannosaurus rex* and not wonder where such a creature came from. Recent finds in Africa and South America have helped bring the origin of dinosaurs into better focus, altering what paleontologists expected. Dinosaurs did not "rule the Earth" from the outset. The reptiles were just one form of many in a strange Triassic world, somewhat marginal creatures that got their shot at the ecological big-time, thanks only to another mass extinction.

Dinosaurs belong to a broader group of creatures called archosaurs, the "ruling reptiles" that also include crocodiles, the flying pterosaurs, and their closest relatives. What's puzzled paleontologists is what sort of archosaur the first dinosaurs evolved from. The task is challenging given how widespread and successful archosaurs were throughout the Triassic. Relatives of crocodiles were abundant and disparate, evolving into a range of forms from four-on-the-floor armored herbivores to bipedal, toothless plant eaters that were effectively dinosaur mimics. The earliest dinosaurs lived right alongside these strange crocs, and so must have arisen from some earlier form of archosaur. The challenge was finding these transitional forms from the older, lesser-known rocks of the Triassic.

Until the beginning of the twenty-first century, most of what paleontologists knew about the Triassic came from rocks between 220 million and 201 million years old that dated from the later part of the period. Paleontologists had found early dinosaurs in places like Petrified Forest National Park, Arizona, and Ischigualasto Provincial Park in Argentina, but many of these dinosaurs were already recognizable members of established dinosaur groups. *Coelophysis*, found in the banded hills of Petrified Forest, was a slender, carnivorous theropod that munched on lizards skittering through ancient conifer forests, while little *Eoraptor*, from Ischigualasto, was a spaniel-size omnivore at the start of the sauropodomorph lineage—the group that would eventually give rise to long-necked, plant-eating giants like *Brachiosaurus*. The ancestors of dinosaurs had to be preserved in even older rocks, formed during earlier parts of the Triassic, that had not been as extensively searched by paleontologists. As it would turn out, paleontologists had been finding fragments of the earliest dinosaurs almost a century ago. They just didn't realize what they had uncovered.

The best current candidate for the oldest dinosaur was found in the 1930s and existed in scientific limbo until 2013. Cambridge University paleontologist Francis Rex Parrington led an expedition to Tanzania's Ruhuhu Basin to look for Triassic fossils in rocks from an earlier part of the period. The team found fossils, as they hoped, including a right upper arm bone and several vertebrae that were placed in the collections of London's Natural History Museum. At the time, the bones seemed to have come from some kind of ancient reptile but little more could be said of its classification. And even though other expeditions collected fossils from the Ruhuhu Basin and placed them in several museums, the fossils were known to only a small number of specialists interested in Triassic

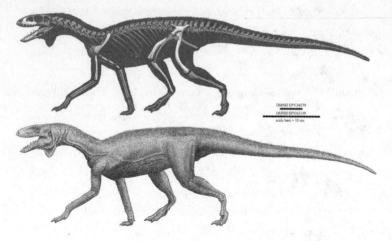

Dinosauriformes such as *Kwanasaurus* have revealed what the first dinosaurs and their immediate ancestors looked like.

life. The arm bone and vertebrae Parrington collected, for example, did not receive even a preliminary description until paleontologist Alan Charig mentioned them as "Specimen 50b" in a 1956 doctoral thesis. Charig also gave the bones an informal name in 1967, but he never formally wrote up the bones as anything special. When Charig passed away in 1997, it seemed the fossils were all but forgotten.

In the early twenty-first century, however, paleontologists began to get interested in the Triassic again. The rise of the dinosaurs was understood only in vague terms and was ripe for a rethink. Previously, paleontologists had assumed that early dinosaurs like *Eoraptor* were anatomically superior to other reptiles and clawed their way to the top through competition. The more experts dug in, however, the more the story of dinosaurian supremacy didn't fit. The Triassic was a strange time, dinosaurs seeming rare and often small compared to crocodile relatives and other creatures of the period. Experts also began turning up creatures that were very much like dinosaurs, yet not dinosaurs at all. *Silesaurus*, named in 2003, was exceptionally dinosaur-like but not quite a dinosaur itself, a slender and lanky reptile about the size of a German Shepherd that walked on all fours and munched beetles. *Silesaurus* belonged to a group that paleontologists called "dinosauriformes," from which the earliest dinosaurs must have emerged. *Silesaurus*, in other words, provided a rough image of what the earliest dinosaurs might have been like—a slender omnivore rather than a terrifying creature that trampled the competition into ecological submission.

The problem was that *Silesaurus* lived too late to be a dinosaur ancestor itself. The reptile was from the Late Triassic, indicating that more ancient dinosaur predecessors lived alongside the earliest dinosaurs for millions of years. The answers paleontologists were looking for had to be found in older parts of the Triassic, and some experts recalled fossils collected by Parrington and others from Tanzania. A multi-institution team of experts set off on new expeditions to the Ruhuhu Basin and had another look at the fossils Charig had outlined decades before. Among them was "Specimen 50b," along with bones from the same species held at another museum, different enough from everything else that had been found to deserve its own name. In 2013, paleontologist Sterling Nesbitt and coauthors named these curious bones *Nyasasaurus parringtoni*, the latest candidate for the earliest dinosaur.

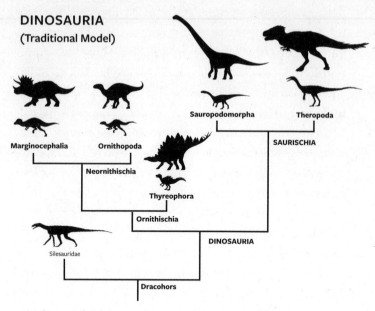

The dinosaur family tree has two main branches that split in the Triassic—the saurischians and ornithischians—and many subgroups in each.

So far, very little of *Nyasasaurus* has been found. Experts have only the upper arm bone and some vertebrae to work from. The anatomical details suggest that *Nyasasaurus* looked something like *Silesaurus*, a slender and svelte reptile with a long neck. What makes *Nyasasaurus* special, though, is that the upper arm bone has a long protruding edge that is seen only among dinosaurs and no other group of Triassic archosaurs. While it's possible that additional finds might alter what paleontologists think *Nyasasaurus* is, at present it is the oldest dinosaur candidate yet uncovered, indicating that dinosaurs had already emerged 233 million years ago. And even if *Nyasasaurus* is eventually displaced, additional finds from Africa and South America have indicated that dinosaurs were already beginning to split into their major lineages 230 million years ago. *Mbiresaurus*, named in 2022, lived at this time and is known from a nearly complete skeleton—enough material to recognize *Mbiresaurus* as one of the earliest sauropodomorphs, the beginning of the lineage that would come to include quadrupedal, long-necked giants such as *Apatosaurus* and *Argentinosaurus*. The emerging picture is that the earliest dinosaurs emerged in the southern part of Pangaea sometime before 230 million years ago and soon began to split into the major dinosaur groups that would proliferate all over the planet.

Sauropodomorphs were one of the first dinosaur groups to evolve, starting as long-necked herbivores that walked on two legs like this *Sarahsaurus*, before giving rise to four-legged long-necks like *Apatosaurus* in the Jurassic.

The challenges in tracking down the earliest dinosaurs raise a basic question: What is a dinosaur? The matter is not so simple as it was in Victorian England, when Richard Owen named the whole group from a small smattering of isolated bones. After all, many different forms of reptiles thrived in the seas, in the air, and on land during the 186 million years of the Mesozoic world—divided into distinguishable Triassic, Jurassic, and Cretaceous periods—many of which were not dinosaurs despite being equally impressive. "Dinosaur" is not a colloquial catchall term, in other words, but a scientific marker as specific as "mammal." To mark the boundaries of what is dinosaurian and what is not, there are two approaches.

To comprehend the big picture of dinosaur evolution, envision a pigeon and the giant, three-horned *Triceratops*. Both are dinosaurs, and each is the embodiment of the two main branches of the dinosaur family tree. The two can be reliably split on the basis of their hips, a fact recognized by paleontologist Harry Govier Seeley in 1888.

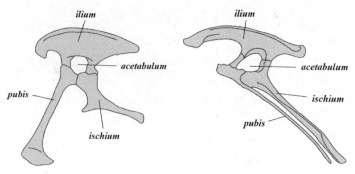

Dinosaurs are divided into one of two groups. Saurischians (left) usually have a more lizard-like hip shape while ornithischians (right) usually have a more birdlike hip shape despite not being close relatives of birds.

The *Triceratops* side of the dinosaur family tree is the ornithischians, or the ironically named "bird-hipped" dinosaurs. Seeley gave the group this name because dinosaurs such as the three-horned *Triceratops*, the armor-plated *Stegosaurus*, the tube-crested *Parasaurolophus*, and more all have the same hip structure—a flange-like ilium above the thin pubis bone which titled backward to fit snugly against a bone called the ischium. The arrangement looks like what's seen in living birds, hence the "bird-hipped" name even though ornithischian dinosaurs were not the ancestors of birds. Instead, ornithischian dinosaurs include the armored thyreophorans like *Ankylosaurus* and *Stegosaurus*, the dome-headed and horned marginocephalians such as *Pachycephalosaurus* and *Styracosaurus*, beaked hadrosaurs including *Hadrosaurus* and *Edmontosaurus*, and related groups, such as the small, tusked heterodontosaurs. Some of these dinosaurs evolved simple feathers, and, so far as we know, every single group was herbivorous.

The bird is among the modern branches of the dinosaur family tree. The pigeon represents the still-living tip of the saurischian, or "lizard-hipped," dinosaurs. Seeley named such dinosaurs saurischians because the hips of dinosaurs such as four-footed *Diplodocus* and theropod *Allosaurus* had the pubis pointing forward, like in lizards, and birdlike species with a backward-pointing pubis, such as *Deinonychus*, had not been found yet. Despite such complications, however, the split works for assessing which dinosaur belongs to which major group, and there are two main branches of the saurischian side of the tree. Sauropodomorphs were saurischian dinosaurs, including small, omnivorous species like *Mbiresaurus* as well as giants like the 110-foot-long (33.5 m), 70-ton (63,500 kg) *Patagotitan*. The other group is known as theropods, the "beast footed" dinosaurs that included carnivores such as the infamous *Tyrannosaurus* as well as omnivorous and plant-eating species like the bizarre, duck-billed and hump-backed *Deinocheirus*.

To delineate Dinosauria, we have to only follow the branches. If we were to start with our *Triceratops* and pigeon and trace our way back toward their last common ancestor—a creature like *Nyasasaurus*—every species more closely related to those two lineages than other reptiles is a dinosaur.

Styracosaurus, for example, is definitely a dinosaur because it's part of the same group—the horned ceratopsids—that *Triceratops* belongs to, but the sleek, ocean-dwelling lizard *Mosasaurus* is not a dinosaur because the reptile had a totally separate ancestry, outside the brackets of Dinosauria.

Naturally, the family tree approach to what is and isn't a dinosaur works only because paleontologists have developed and revised a dinosaur family tree over decades. It's useful for working out whether a particular animal should be considered a dinosaur or not on the basis of its relatives, but the approach doesn't really tell us what specifically makes a dinosaur a dinosaur. For that, we need to get into some anatomical particulars.

Paleontologists, like many biologists, work out relationships between living things based upon shared characteristics. The more unique characteristics two living things share, the more closely related they are. Just like duck-billed platypus for example, our species has a specialized set of tiny bones in our inner ear, furry body coverings, and the ability to make milk, which are all traits shared in common by mammals. But the duck-billed platypus still lays eggs while our species gives live birth to young that are internally nourished via a placenta during early development, and these traits designate us as placental mammals while the waddling platypus belongs to a group called monotremes. And by using these sorts of biological comparisons, paleontologists have been able to come up with a short list of features that dinosaurs share to the exclusion of any other reptiles.

During the twentieth century, it seemed relatively easy to tell dinosaurs apart from other reptiles that lived during the Triassic, Jurassic, and Cretaceous periods. All dinosaurs have column-like limbs held in an upright position below the body, different from the sprawling posture of lizards or the slightly bent position seen in crocodiles and alligators today. Dinosaurs also have simple, hinge-like ankles in which the bones of the lower leg and the bones of the upper foot make a relatively straight, clean gap, whereas the ankles of other archosaurs have a more complex arrangement. For the most part, these recognizable traits in the limbs can easily help museum-goers tell dinosaurs apart from animals that are just dinosaur-like. But it wasn't so simple during the Triassic, when dinosaurs were just getting their start. Some crocodile relatives evolved upright limb postures just like dinosaurs did. One particular group of crocodile relatives called poposaurids included both carnivorous and omnivorous species that stood and walked just like theropod *Coelophysis* and other early dinosaurs. To really tell dinosaurs apart, paleontologists had to dig deeper into reptilian anatomy.

A massive survey of the dinosaur family tree published in 2011 identified twelve distinct dinosaur traits that can reliably distinguish the terrible lizards—provided the right bones are found. While many reptiles have an opening in the rear top of their skull called the supratemporal fenestra, for example, only dinosaurs have a depression just in front of it called a supratemporal fossa. And while other animals have a flange on their upper arm bone, or humerus, in dinosaurs the flange is situated at about a third of the way down the bone or more. In addition to other subtle features in the ankles, spine, legs, hips, and so on, the shared characteristics can help act as anatomical tiebreakers when it's unclear whether a fossil reptile is a dinosaur or merely similar in form.

As paleontologists have followed these skeletal hallmarks to identify the earliest dinosaurs, the story of how dinosaurs became so successful has turned into an underdog story. From the time of *Nyasasaurus* until the end of the Triassic 201 million years ago, dinosaurs were usually not the most numerous, diverse, largest, or even most fearsome creatures of their prehistoric habitats. The Triassic world was one of vast conifer forests and floodplains crisscrossed by streams. In these habitats, you would be more likely to spot prehistoric crocodile relatives than early dinosaurs. On land, at least, what paleontologists call the "crocodile-line" reptiles—technically called pseudosuchians—evolved into a vast array of shapes, sizes, and habits while dinosaurs were on the ecological sidelines. The spiked aetosaurs shuffled through forests, knife-toothed carnivores such as *Fasolasuchus* chased after prey, greyhound-like spehosuchians scampered through the ferns, and bipedal herbivores such as *Effigia* presaged forms that dinosaurs would copy millions of years later. Even though the Triassic is often called the "Dawn of the Dinosaurs" because of the famous grouping's emergence, if you were to visit the Triassic, it would look more like a Dawn of the Crocodiles.

Dinosaurs were different from the crocodiles and other reptiles of the time. A tiny creature found in Madagascar helps highlight the critical differences. Described by paleontologist Christian Kammerer and colleagues in 2020, *Kongonaphon kely* isn't a dinosaur so much as a near-dinosaur ancestor—a creature close to the forerunners of the first dinosaurs. *Kongonaphon* would have been only about 4 inches (10 cm) tall and its teeth seem to be worn down from insect shells, a diet held in common with quadrupedal, dinosaur-like reptiles called "silesaurids." Feeding on such energy-rich food, as well as the microscopic structure of the reptile's bones, hint that it maintained a warm and relatively constant body temperature, likely assisted by a coat of insulating fuzz seen in some small dinosaurs. In other words, dinosaurs inherited insulating coats of early feathers, warm-running physiologies, and their specialized posture from tiny ancestors, evolutionary gifts that would make all the difference when life on Earth suffered another terrible shock.

Toward the end of the Triassic, volcanoes in the center of Pangaea began to ooze and suppurate vast amounts of molten rock once more. Geologists know this area as the Central Atlantic Magmatic Province, or CAMP, and remnants of these rocks can still be found in South America, Africa, and North America, including New York's famous Palisades. The outpouring wasn't constant. The eruptions pulsed over a period of 600,000 years, with lava flowing over more than 4.2 million square miles of the planet and pouring incredible amounts of climate-altering compounds into the air. The Earth did not just warm, but swung between hothouse conditions and volcanic winters capable of producing ice fields near Earth's poles. The planet's climate and seasons were thrown into flux, putting new pressures on creatures that had evolved in much more stable and predictable conditions.

The ecological turbulence was too much for many creatures to keep up with. Ancient animals such as enormous, salamander-like amphibians called metoposaurs entirely vanished at the end of the Triassic, as well as the superficially crocodile-like phytosaurs that once filled Triassic waterways. Crocodiles and their relatives didn't fare very well, either. The vast majority of their lineages went extinct, too, leaving only a few small species that would carry along the pseudosuchian legacy into the following Jurassic period and require

The double-crested *Dilophosaurus* lived in the Early Jurassic as dinosaurs thrived in the aftermath of the Triassic-Jurassic mass extinction.

crocodiles to diversify anew. And this is to say nothing of the extinctions in the seas, complex reef ecosystems once again reduced to communities of straggling survivors. Dinosaurs, however, appear to have walked through all of this turmoil unfazed. To date, there is no evidence that there was a mass extinction of dinosaurs at the end of the Triassic. Theropod, sauropodomorph, and ornithischian dinosaurs all split from each other and began to diversify during the Triassic and all these groups are still present in the earliest Jurassic rocks.

Dinosaurs inherited a Jurassic world freed from their closest competition for food and habitat. Paleontologists are still investigating what made the difference, especially given some of the similarities in diet and ecology to the crocodiles that were tipped over into extinction. The leading hypothesis is that having warm, relatively constant body temperatures and insulating coats allowed dinosaurs to inhabit a much broader range of habitats than "cold-blooded" creatures whose body temperatures followed those of their environment. The difference meant that not only did dinosaurs spread widely around Pangaea and into a vast range of habitats—a basic kind of insurance against total extinction—but they also stayed warm during volcanic winters that reptiles, giant amphibians, and other creatures could not handle. Unique traits that dinosaur ancestors evolved at small size might have saved them during the fourth of Earth's Big Five mass extinctions, allowing dinosaurs to walk into an era nearly devoid of the other reptiles that constrained the available evolutionary pathways. The Jurassic period was when dinosaurian evolution truly began to bloom, setting the stage for the emergence of reptilian wonders not seen before or since.

CHAPTER 2

Eggs and Nests

Every dinosaur has started life the same way. Each and every dinosaur that has ever existed hatched out of an egg. From little *Eoraptor* in the Triassic and *Triceratops* at the end of the Cretaceous to every bird alive today, all dinosaurs greeted the world by breaking out of a shell. The reproductive happenstance is one of the central secrets of dinosaurian success. Despite the fact that we're still awed by the size, strangeness, and apparent ferocity of many dinosaurs, it's the fact dinosaurs laid eggs that allowed them to so quickly spread through the Mesozoic world—the entirety of the era marked by dinosaurs—and thrive in so many different forms. The story of dinosaurs is, when you boil it down, about eggs.

What we now know about dinosaur eggs and nests has been carefully assembled from some of the rarest fossils in the entire dinosaurian record. It's a wonder that we know anything at all about these parts of dinosaur lives. Eggs and nests were only ephemeral parts of any dinosaur's life story. Eggs remained nestled in the same habitat for months, sometimes watched over by parents and other times left to chance under a blanket of sediment and vegetation. Often, those eggs eventually busted apart and fell away to release tiny dinosaurs that met an entire lifetime of challenges. But if every dinosaur egg hatched, we wouldn't really know anything about them. Ancient misfortune has allowed us to at least outline how many dinosaur species met the world.

For dinosaur eggs to be preserved, something had to go wrong. The hope contained in each developing egg was sometimes lost as prehistoric streams overflowed their banks to spill sediment over the landscape, ash cascaded down from volcanic eruptions, vast sandy dunes collapsed onto dinosaurs tending their nests, or some other stroke of bad luck buried the eggs and wrapped them up into the fossil record. Of course, more buried eggs were likely lost entirely than became fossils. Whether soft or hard, eggs are by their very nature thin-walled and sometimes fragile. Many broke down and decayed before the fossilization process could begin. It speaks to the sheer volume of eggs dinosaurs produced during the Mesozoic that paleontologists have as many of these rare fossils as they do.

The first recorded discovery of dinosaur eggs was made among the foothills of Europe's Pyrenees Mountains in 1859. The Catholic priest Jean-Jacques Pouech was curious about the rocks and fossils found in the vicinity of his seminary in southern France, geological remnants of the area about seventy million years earlier during the Late Cretaceous. Among the curiosities he stumbled upon were strange fragments that he initially thought were armor plates of some large reptile. On closer inspection, though, the consistency in the fossil surfaces, their structure, and their curvature led Pouech to recognize the fossils as eggshell, what he estimated as "at least four times the volume of ostrich eggs." Whatever had left the eggs behind, it was far larger than those laid by any modern creature.

Pouech, like other naturalists puzzling over early dinosaur discoveries, thought that the eggs might have been laid by enormous birds of some bygone age. He had no concept of what a dinosaur was, the term having been coined only seventeen years earlier. Naturalists were familiar with dinosaurs, but the bizarre reptiles were not yet well known outside scientific circles. Even when Pouech showed the eggshell fossils to Paris' Muséum National d'Histoire Naturelle, naturalists there disagreed and denied that the fossils came from eggs. Pouech eventually moved on to other things, his historic finds forgotten until 1989 when paleontologists relocated his collection and confirmed the identity of the eggshells. Precisely what species of dinosaur laid them is unknown, but, based on similarities in structure to dinosaur eggs found elsewhere and other fossils in southern France, Pouech's eggs were probably laid by a long-necked sauropod dinosaur like *Hypselosaurus*.

Paleontologists have found dinosaur eggs and nests all over the world. Among those most commonly found were the parrotlike oviraptorosaurs' oblong eggs, which were laid in pairs in nests tended by the parent dinosaurs.

EGGS AND NESTS

Their eggs were not quite so massive as Pouech guessed, but they were still laid en masse in vast nesting grounds where embryos would gestate for several months before busting out of their shells.

The ancient herbivore wasn't unique in laying eggs and creating nests. The spread of evidence indicates that all dinosaurs laid eggs, an inheritance from their reptilian ancestors that has never changed. Even though other reptiles—both living and fossil—have evolved different ways of reproducing, including live birth, dinosaurs stuck with eggs and nests as they spread across the planet. In many ways, the different shapes and sizes dinosaurs would come to embody can be attributed to the consequences of reproducing via eggs.

The earliest dinosaurs were egg-layers, just like their ancestors. And even though cynodont and dicynodont protomammals that survived into the Triassic laid eggs, too, it was egg-laying paired with dinosaur life history that gave the reptiles some evolutionary acceleration. The microscopic details of their bone tissue indicate that Triassic dinosaurs were very active animals that grew quickly, a profile that would generally be conserved by their descendants. But whereas some animals begin reproducing only when their growth slows and they reach full adult body size, dinosaurs evolved to do something different. A peculiar tissue found in the bones of some Jurassic and Cretaceous dinosaurs indicates that dinosaurs began making nests and laying eggs long before they reached their maximum body size. The clues are directly tied to egg-laying, hidden inside dinosaur bones in the form of a specific tissue called medullary bone.

The discovery of medullary bone didn't come from fossil dinosaurs, but from studies of birds. When birds are about to lay their eggs, calcium-rich bone tissue builds up inside the central cavity of some of their long bones. The temporary buildup in extra tissue provides raw materials for each bird embryo to be enclosed in a protective shell, the essential calcium needed to form resilient armor. A *Tyrannosaurus rex* skeleton discovered in 2000 provided the first evidence that non-avian dinosaurs underwent the same change associated with egg-laying.

Nicknamed B-rex after Bob, the worker who noticed one of her bones protruding from a rock wall, the dinosaur was not one of the more complete *T. rex* specimens known. Paleontologists recovered only about 37 percent of the tyrannosaur's skeleton. Among the excavated B-rex bones, however, were both thigh bones, or femora. The long bones are like time capsules of the dinosaur's life. Details of the microscopic bone tissue within can help paleontologists estimate how fast a dinosaur grew, how old it is, and more. When biologist Mary Higby Schweitzer and colleagues looked at the slides of B-rex femur sections, however, they found additional bone tissue within what should have been a central hollow. Upon comparison with other bone tissue types, the researchers found a close resemblance to the medullary bone of birds. If B-rex wasn't already laying eggs when they perished, they were about to.

The finding revealed several essential traits of B-rex, as well as non-avian dinosaurs more broadly. B-rex was a female dinosaur, the first dinosaur to ever be confidently sexed from direct evidence. The fact that B-rex did not look significantly different from other *T. rex* specimens lacking medullary bone, too, indicated that the tyrannosaur was not sexually dimorphic: At least so far as their skeletons go, *T. rex* of different sexes look the same. On top of that, an additional analysis by paleontologists Andrew Lee and Sarah Werning found the thigh bones of B-rex contained enough growth rings inside for experts to estimate the dinosaur's age at death. Dinosaurs that lived in environments with distinct wet and dry seasons grew during the wetter months when food was more plentiful and slowed or ceased their growths during the more stressful dry seasons, similar to how trees grow and rest according to the shifting conditions of the seasons. Paleontologists call these lines of arrested growth, or LAGs. By counting these rings and the nature of the dinosaur's bone tissue, Lee and Werning found that B-rex was about eighteen years old when they died. The oldest known *T. rex* had perished at about thirty, and B-rex was still growing at the time of death. Taken all together, the hidden dinosaur details revealed that *T. rex* were laying eggs and making nests long before they reached full adult size. Medullary bone found in two additional dinosaur species— the herbivorous, beaked *Tenontosaurus* from 125 million years ago and the carnivorous, horned *Allosaurus* from 150 million years ago—followed the same pattern. Dinosaurs were mating and laying eggs early in their lives.

Laying multiple eggs early in life allowed the first dinosaurs to spread through the Triassic world, outpacing protomammals but not quite matching the array of crocodile relatives. After the end-Triassic extinction, dinosaurs retained the same reproductive profile and kept creating clutches of eggs relatively young. As dinosaurs began to evolve into larger sizes and became more distinctive, the first truly large dinosaur carnivores began to emerge. It was at that point that breeding young became a way for dinosaurs to avoid being eaten into extinction.

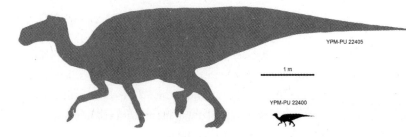

All non-avian dinosaurs hatched from eggs and were born smaller than their full-size parents. Studies of baby *Maiasaura* indicate that dinosaur babies were vulnerable to predators, drought, and other dangers until they were about one year old.

The rarity of yearling and juvenile dinosaurs may be because carnivores, much like predators today, specifically targeted the small, relatively naïve young dinosaurs that could not successfully ward off the hungry jaws and claws of the time. Even the largest dinosaurs started small and would have been vulnerable to predators, on top of risks from disease, drought, and other hazards of the Mesozoic world. In fact, the study of one dinosaur bone bed found that nearly 90 percent of infant dinosaurs died during their first year. Dinosaur hatchlings faced a litany of dangers and were much more likely to perish in their first year than survive it. Likewise, paleontologists have found very few old dinosaurs. When dinosaurs stopped growing they formed a line inside their long bones called an external fundamental system, or EFS, that indicates a late life slowing of growth. Not many dinosaurs with an EFS are known. Most dinosaur skeletons that paleontologists find represent animals that were fortunate enough to survive the critical first year but never finished growing before they perished. Despite the deadly dangers dinosaurs faced, however, laying many eggs early in life allowed the reptiles to thrive on an ever-changing planet. The greater the number of eggs laid each breeding season, the greater the chances that a small fraction would survive and become old enough to create their own nests, as well as increasing the chances an individual dinosaur would pass on their traits, and those of their mate. In a volatile world, dinosaurs kept up by reproducing fast.

Our knowledge of such nests is only about a century old. Paleontologists knew practically nothing about dinosaur nests until the 1920s. Repeated expeditions by New York's American Museum of Natural History to Mongolia in that decade turned up eggs and nests, particularly from the dog-size horned dinosaur *Protoceratops*, and even dinosaurs thought to be dedicated egg-eaters. Paleontologists found the bones of a parrotlike dinosaur in proximity to fossil eggshells. The theropod was named *Oviraptor*, or "egg thief," and decades went by before a fresh look at the evidence revealed that the dinosaur was not stealing eggs so much as watching over them. Continued searches of the Gobi Desert have turned up many more oviraptorosaur fossils, including some parent dinosaurs sitting on nests of delicately arranged eggs with their arms outstretched over them. Dinosaurs not only made nests, but some species protected them. The "terrible lizards" evolved their own forms of parental care.

In the century since the Gobi Desert expeditions, paleontologists have uncovered dinosaur eggs and nests of various different groups and species. The fossil affectionately named "Baby Louie," for example, is an embryo of an *Oviraptor*-like dinosaur found among a nest of oval-shaped eggs arranged in pairs. Other eggs are found as shell fragments or in isolation. In 1989, paleontologists found a single, pathological egg from the Jurassic rock of Utah's Cleveland-Lloyd Dinosaur Quarry. The shell had more than one layer, a common abnormality seen among stressed birds. It wasn't until decades later that the discovery of a Jurassic nest site in Colorado containing eggs and embryos allowed the identity of the singular egg to be determined. The embryonic bones were *Allosaurus*, and the eggshell at the nest site matched the singular Cleveland-Lloyd egg, nestled in a bone bed of multiple *Allosaurus*. Nesting sites made by other dinosaurs have also turned up here and there around the world. Long-necked sauropod dinosaurs congregated at nesting grounds throughout their history; Cretaceous fossil sites in Patagonia and India contain dozens of nests. One particular site in Argentina, found in 2010, contained at least eighty clutches of round eggs numbering between three and a dozen each. All of the nests were placed about 3 to 10 feet (1 to 3 m) away from hydrothermal features, like those found in Yellowstone National Park's geyser basins. The dinosaurs were making their nests near thermal pools that likely bubbled and steamed. Whether the egg-laying sauropods knew why this spot was different or not is unclear, but the water in the soil and radiant heat from the hydrothermal features helped keep the eggs warm and moist. Paleontologists expect that sauropod dinosaurs laid their eggs and simply left them, much like sea turtles do with their nests, and so the hydrothermal vents effectively babysat the dinosaur eggs during their months-long incubation.

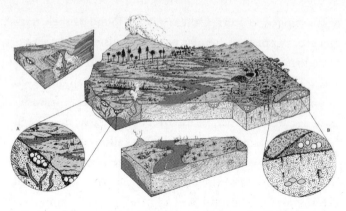

Sauropod dinosaurs nested in different habitats, such as hydrothermal fields (A) or mounds made in low-lying floodplains (B).

Not only were some dinosaurs specific about where to make their nests, but some species returned to the same spots year after year to lay their eggs. Biologists know this phenomenon as site fidelity, where there is a particular spot individuals of a species return to as they reproduce. One of the oldest dinosaur nesting sites known was found in 1976 among the roughly two-hundred-million-year-old rocks of Golden Gate Highlands National Park in South Africa. New excavations of the site in 2006 not only found a total of ten egg clutches of the same dinosaur, an early sauropodomorph called *Massospondylus*, but also that the nests were buried at different layers within the red rock. At least two *Massospondylus* nested in the same place at least four times between several seasons, leaving as many as thirty-four eggs spaced out in a single layer. Some bird and crocodilian species nest the same way today, coming back to the same spot in a communal nesting ground over the years.

Just as experts can look into the microscopic details of dinosaur bones and even their geochemical makeup, so have researchers looked into the microscopic structure and bio-chemical nature of dinosaur eggs. The subfield has been able to distinguish the eggs of different species by the microscopic details of the shell layers and even determine whether eggs were covered in the nest or exposed to the air. Paleontologists have even been able to work out how long little dinosaurs gestated inside their shells. So far as paleontologists have been able to estimate, dinosaur eggs were incubated for between 2.8 and 5.8 months. The rate is relatively slow compared to how long birds need to incubate, which is between 11 days and 2.8 months. The extended incubation period among non-avian dinosaurs increased the risks for the eggs. Predators of all sorts took advantage. In western Colorado, near a nesting site for a small, herbivorous dinosaur that walked on two legs called *Dryosaurus*, experts have found a tiny terrestrial crocodile called *Fruitachampsa* that almost certainly dined on eggs and hatchlings. Paleontologists have found large fossil snakes among fossil nesting grounds, such as *Sanajeh* found within a sauropod nesting ground in India, leaving little doubt that dinosaur eggs were surely seen as meals by various ancient creatures. The challenges of survival began before baby dinosaurs even emerged from their shells. It seems dinosaurs evolved two different responses to these risks.

Some dinosaurs didn't look after their offspring at all. Nesting grounds from the large sauropod dinosaurs show no evidence of parental care or that hatchlings spent a great deal of time in the nest. Dinosaurs such as *Apatosaurus* and the giant quadruped titanosaurs laid large numbers of eggs, often in communal nesting grounds, and simply left them. Paleontologists refer to this as a "lay 'em and leave 'em" strategy. Which babies survived was simply a numbers game, so many hatching at once it was near impossible for predators to eat them all. But various other dinosaurs looked after their eggs while the developing babies were vulnerable. Nests from the duck-billed dinosaur *Maiasaura* initially discovered by Marion Brandvold indicate that the babies stayed in or near the nest for at least a few months after hatching as adults provisioned them with food. The care offered some extra protection for the little ones, especially because predators seemed to nest near the large herbivores. Paleontologists have found small, raptor-like dinosaurs sitting on their nests near the *Maiasaura* nesting grounds, the adult carnivores perhaps picking off baby *Maiasaura* to feed to their own hatchlings. Just like modern birds and crocodylians, then, dinosaurs probably had a range of different parental care styles, from attentive hatchling-rearing to leaving the eggs to chance.

Some dinosaurs were attentive parents who looked after their nests and offspring.

When it came time to hatch, baby dinosaurs had different ways of breaking out of their eggs. Theropod dinosaurs likely kicked out of their eggshells the same way birds like emus do today, pressing with their feet to crack the eggshell from the inside out. Some sauropod dinosaurs might have pushed their way out with specialized structures on their tiny snouts. A titanosaur embryo described in 2020 had a pointed, horn-like structure on its face. The projection was not the same as the temporary egg tooth seen among some reptiles and birds, but may have served a similar purpose in helping some of these dinosaurs push their way out of their eggs. How armored dinosaurs, horned dinosaurs, and others emerged, however, is as yet unknown. Dinosaur eggs and embryos are so comparatively rare that almost any finding helps clarify how baby dinosaurs met the world.

Paleontologists are still just beginning to understand the sheer variety of different ways dinosaurs nested, looked after their offspring, and hatched. There was no single way to make a baby dinosaur or create a good nest. Dinosaurs eggs, nests, and parental behaviors were likely as disparate as all the different forms of dinosaurs, and all the different lifestyles and environments requiring a varied array of reproductive strategies. Whatever dinosaurs did, though, their efforts worked spectacularly. Every dinosaur skeleton experts uncover is an animal that began life hatching from an egg, laid by a parent who also hatched from an egg and managed to survive long enough to make their own nest. Season by season, and egg by egg, dinosaurs thrived.

CHAPTER 3

Growing Up Dinosaur

When paleontologists name a new dinosaur, they hope the name will stick. Experts spend a great deal of time assessing and comparing bones, determining which salient traits mark a fossil as a member of a distinct species, and then proposing the fossil belongs to a particular species with a consistent name. What we know as the "tree of life" is founded in such research and distinctions. Often, especially when the fossil is relatively complete, experts can feel confident that they've identified a true species that can readily be distinguished from others. The dinosaur *Triceratops*, for example, is immediately recognized by its two brow horns, single nose horn, and a frill of solid bone behind its head, a set of easy-to-spot characteristics different from other horned dinosaurs like *Styracosaurus*, with a long nose horn and long spikes jutting from its frill, and *Pentaceratops*, named "five-horned face" due to its prominent horns on the nose, brows, and cheeks. But paleontologists are always working from an incomplete fossil record, not just missing parts of individual animals but representations of different life stages, too. Ideally we'd know every dinosaur species from a few dozen complete skeletons of various ages, eggs, footprints, skin impressions, and other clues, creating as complete a picture of the dinosaur as the fossil record can allow. Paleontologists would love to have as many specimens of *Stegosaurus* as, for example, ravens from a particular population, allowing their variations and life histories to be teased out in deep detail. Such ideals are almost impossible to meet. The fragmentary nature of the fossil record requires constant comparison and revision to work out dinosaur identity and relationships, and sometimes what seemed like a distinctly new dinosaur turns out to be one we already know.

A century ago, paleontologists recognized only a fraction of the dinosaur species we know of today. What experts and the public knew of dinosaurs was mostly represented by Late Jurassic classics such as *Diplodocus* and *Stegosaurus*, along with favorites from the Late Cretaceous such as *Tyrannosaurus* and *Triceratops*. With a smattering of exceptions, experts mostly knew dinosaurs from a few million years—out of a Mesozoic span that stretched more than 180 million. Almost any skull or skeleton that came out of the ground and looked immediately distinct received a new name. Experts were not yet thinking of dinosaurs as living animals that hatched, grew up, and took on different traits as they aged. In fact, because of misunderstandings about the way modern reptiles grow up, paleontologists thought that baby dinosaurs looked just like miniaturized versions of their parents. Small dinosaurs that looked distinct were considered small or dwarfed species, with very few skeletons recognized as those of juveniles. On top of that, even adult dinosaurs were sometimes named from parts of their skeletons that didn't match up to comparable bones from related species, so a paleontologist might come up with a new name for a skull and a partial body without recognizing that both actually represent the same species. Given that the same animal might be given several names by different experts, paleontologists quickly created a taxonomic tangle. The same animal might be bestowed different names several times over, each expert having their own preference for the proper title. The duck-billed dinosaur we now call *Edmontosaurus*, for instance, had variously been named *Anatosaurus*, *Anatotitan*, *Claosaurus*, and others before experts realized the fossils were just *Edmontosaurus* by another name.

The way dinosaurs grew up has certainly played into these complications. Small dinosaurs once thought to be distinct have often turned out to be juveniles of dinosaurs first known from adult specimens. A puzzling dinosaur found in Canada is an illustrative case. Even though many museums built their early dinosaur collections from Jurassic rocks, by the 1910s experts were turning their attention to the roughly seventy-five-million-year-old rocks of Alberta. The colorful, banded deposits were not only full of dinosaur bones, but paleontologists often found near-complete, articulated skeletons of dinosaurs such as the carnivorous tyrannosaur *Gorgosaurus* and the horned ceratopsian *Centrosaurus*. So many dinosaurs seemed to be spilling out of the ground that sometimes paleontologists collected only the skull, leaving the rest of the skeleton behind to be rediscovered by later generations of experts. Among these impressive finds was the skeleton of a small duck-billed hadrosaur named *Procheneosaurus*. The little honker was about the size of a cow and lacked the elaborate crests of other hadrosaurs from the surrounding area, like the tube-crested *Parasaurolophus* or the helmet-headed *Corythosaurus*. The dinosaur seemed to be a diminutive duck-bill that lived alongside its enormous relatives among the conifer forests and floodplains.

Almost immediately, the true identity of *Procheneosaurus* was brought into question. The process by which the dinosaur was named didn't precisely follow the nomenclature rules for new species, and so *Tetragonosaurus* was proposed as the little herbivore's proper title. But then a petition to the organization that oversees how living things are named discarded *Tetragonosaurus* in favor of the original *Procheneosaurus*. And so it stayed for three decades, until 1975. In that year, paleontologist Peter Dodson recognized that fossils of "*Procheneosaurus*" and other small hadrosaurs were actually juvenile specimens of the crested hadrosaur *Lambeosaurus*, which had a hollow, hatchet-like ornament on its head. No one had recognized the true identity of the little dinosaur because no one had yet recognized how much dinosaurs changed as they grew up.

Early twentieth century paleontologists probably would have recognized the little *Lambeosaurus* if the adolescent

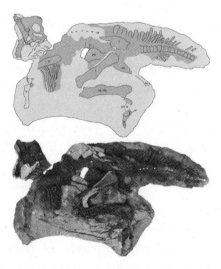

"Joe" the baby *Parasaurolophus* was about one year old when they died, showing only the beginnings of the crest seen in the adults.

dinosaurs had flashy crests. But as recent finds have indicated, hadrosaurs took time to grow into their headgear.

In 2013, paleontologist Andy Farke and colleagues described a baby *Parasaurolophus* from the rocks of southern Utah of about the same geologic age as *Lambeosaurus* in Alberta. As an adult, *Parasaurolophus* is immediately recognizable. The shovel-beaked dinosaur bears a hollow, tubular crest jutting from the back of its skull. (The hollows inside help act as resonating chambers, which allowed the dinosaur to make low, booming calls.) While some bones were missing and the preserved parts of the skeleton were weathered by the soggy conditions under which the dinosaur was buried among ancient swamps, the complete skull was preserved with the body. The head of the little dinosaur nicknamed "Joe" showed a small bump just in front of the eyes instead of the long crest seen in the adults. Other subtle traits on the rest of the skeleton legitimized its identification as a *Parasaurolophus*. And when experts looked to the bone microstructure to estimate the dinosaur's age, they didn't find any growth rings at all. Joe may have been as young as a year old, a juvenile dinosaur with only the barest beginnings of the crest they would have grown if they'd survived to adulthood.

By scanning the small dinosaur's skull and comparing it to those of adults, Farke and colleagues were able to detect how the hadrosaur's crest expanded as it aged. The main airway inside the crest expanded backward, over the rear of the skull. The dinosaur's crest started forming when it was about a third of its adult size, which suggests that the crests were important to the dinosaur's social life. In addition to producing distinct calls, crests were gaudy visual structures that helped *Parasaurolophus* recognize each other and perhaps even choose mates. They may have even been brightly colored, a biological billboard for each individual animal. Little dinosaurs were born big-eyed and cute, but rapidly changed as they grew up, gaining flashy features during their equivalent of teenage growth spurts. Dinosaurs started mating early, well before their skeletons reached adult size, and Joe seems to fit the same pattern.

Hadrosaurs are far from the only dinosaurs to be reinterpreted and reorganized as hatchlings, juveniles, subadults, and adults. Horned dinosaurs, tyrannosaurs, sauropods, and more have species that were folded together because supposed differences turned out to be changes with growth, or what biologists know as ontogeny. In 1979, for example, tiny dinosaur fossils from the Early Jurassic of Argentina were named *Mussaurus* for being almost mouse-size. The dinosaurs had large eyes and had bodies best suited to walking on all fours. It wasn't until 2013 that paleontologists Alejandro Otero and Diego Pol described the adults, sauropodomorph dinosaurs with long necks, large claws, and that walked on two legs. The "mouse lizard" grew up to be 26 feet (8 m) long and weigh 1.5 tons (1,300 kg), a case of unintentional paleontological irony. Dinosaurs both new and old are subject to these continued revisions. Among horned dinosaurs, the supposedly frilled small horned *Brachyceratops* named in 1914 turned out to be a juvenile of the many-horned *Styracosaurus*, and the small horned "bone-headed" *Dracorex* turned out to be a young *Pachycephalosaurus* before its knobbly skull grew a prominent dome. The orientation of horns on a dinosaur's skull, the size of their crests, whether they walked on four legs or two, and much more could all change between hatchling and the rare adulthood every dinosaur lived to reach. Assessing how old a dinosaur was when it died and what stage of life it was in has become a standard part of new dinosaur descriptions in order to forestall additional misidentifications.

Properly identifying young dinosaurs has become a critical paleontological task for several reasons. The most apparent is getting a proper count of how many dinosaur species were present in an ecosystem at one time. If a single species is misidentified as several, then paleontologists might accidentally overestimate the number of species an ecosystem hosted and how the dinosaur tree of life changed through time. Working out which dinosaur species were present at one time is part of the basic information the rest of our fossil knowledge flows from. As experts have become even more adept at these identifications, however, an unexpected trend has emerged. Young dinosaurs didn't just look different from adults. Their anatomy required them to live in different ways, especially when it came to what they ate.

Baby *Diplodocus* grew fast and principally fed on lower-growing, more calorie-rich plant foods than their parents. The rounded shape of their muzzles is best suited for browsing or selecting particular foods, rather than broadly grazing like their parents.

Consider the iconic Jurassic dinosaur *Diplodocus*, an animal that's not only known from multiple adult skeletons but a handful of juvenile fossils, too. Adult *Diplodocus* could grow to more than 100 feet (30.5 m) in length and weigh more than 40 tons (36,000 kg). As an adult, the dinosaur could stand in one place and graze, holding its head low to the ground while cropping the Jurassic vegetation of ferns and horsetails with its pencil-like teeth and swallowing the green bulk whole. The dinosaur's skull is the key clue to this feeding behavior. The front of an adult *Diplodocus* muzzle is flat and squared off, resembling that of a cow and other mammals that graze on low-growing food today. In fact, zoologists have noticed a significant split between grazing herbivores that eat large amounts of low-growing food that's low in nutritional quality and those that more selectively pluck leaves, fruit, and other green food, known as browsers. While grazer muzzles are squared off, those of browsers are more rounded as befits their habit of picking a smaller volume of higher-quality food.

When paleontologist John Whitlock and colleagues described a long-forgotten juvenile *Diplodocus* in 2010, the little dinosaur did not have the square muzzle of its parents. Instead, the muzzle was narrower and rounder: The little one had the face of a browser. Rather than planting their feet in one place and vacuuming up whatever plant food was in reach, it seems that the smaller *Diplodocus* wandered through the Jurassic groves picking and swallowing particular plant food that it needed to fuel its rapidly growing body. Laboratory tests on ancient plants such as horsetails and ginkgoes have revealed that even in the Jurassic, there were calorie-rich plants available to *Diplodocus* and other plant-eating giants of the time. *Diplodocus* had to grow fast. They hatched out of relatively small eggs, meaning baby *Diplodocus* would have been bite-size for the likes of the sharp-toothed theropods *Allosaurus* and *Torvosaurus* stalking the same habitats. The rarity of juvenile dinosaur fossils from the same rocks likely speaks to how much carnivores relied on annual gluts of small sauropods to munch on. Survival required becoming too big and troublesome to attack, which led *Diplodocus* to evolve a rapid growth rate. A skull suited to browsing and

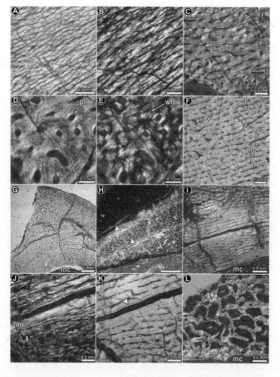

Paleontologists look at microscopic clues for the dinosaurs' ages when they died and were preserved. Under the microscope, fast-growing bone often has a messier, lesser-organized appearance, which indicates that the dinosaur was still growing when they perished.

being selective would have helped young *Diplodocus* get the higher-calorie food they needed, their skulls broadening as they aged thanks to trade-offs between body size and food requirements seen in animals even today. For large animals, such as adult elephants or *Diplodocus*, the overall volume of food is more important than quality. They are not growing as fast and so can subsist of food that's not as nutritious but is more broadly available. Smaller animals with higher metabolic rates, however, need to acquire more energy faster, and so little *Diplodocus* was a pickier eater than the adults in the same ancient environments. The change in skull shape was tied to a shift in diet and natural history, young *Diplodocus* living almost like a different species compared to their adult counterparts.

Some dinosaurs took these anatomical shifts to extremes, a single species taking up ecological roles that would normally support several. There is no better example than *Tyrannosaurus rex* itself, all tied to the fate of one of the most controversial fossils in dinosaur paleontology.

In 1946, paleontologist Charles Gilmore published a short note on a striking tyrannosaur skull uncovered in the Cretaceous rocks of Montana. Despite some damage and distortion, the skull was remarkably intact and was clearly from a small tyrannosaur with large, round eye sockets. Overall, Gilmore proposed, the fossil looked similar to another tyrannosaur found in Alberta called *Gorgosaurus*, but lacked some of the ornamentation of the northern species. He named it *Gorgosaurus lancensis*, what appeared to be a small tyrannosaur distinct from the giant *Tyrannosaurus rex*.

During the following years, experts effectively forgot about the small fossil. It was one of a long list of partially known dinosaurs that early twentieth-century paleontologists had named but whose identity remained dubious. It wasn't until 1988 that paleontologist Phil Currie and colleagues took another look and proposed that the skull Gilmore described was something even more unusual. The skull had come from rocks the same age as *T. rex* itself, between sixty-eight million and sixty-six million years old, but seemed to represent an adult animal that was about the size of a polar bear. The fossil appeared to represent a small tyrant that lived alongside the much larger *T. rex*, and so the paleontologists renamed the skull *Nanotyrannus lancensis*.

As experts continued to compare little *Nanotyrannus* to other tyrannosaurs, however, the fossil's identity opened up a new controversy over how many tyrannosaurs lived in North America at the very close of the Cretaceous. The texture of the bone and the large eyes were more consistent with those of a juvenile tyrannosaur. Dinosaur skull bones tend to fuse as the animals grow up, too, but the *Nanotyrannus* skull didn't show the expected bone fusion for an adult animal. On top of that, repeated explorations of the Late Cretaceous rocks of Montana and surrounding states had turned up multiple adult *T. rex* skeletons but seemingly no juveniles. Small tyrannosaur fossils from these rocks had traditionally been given different names, yet none seemed substantially different from *T. rex* itself. The *Nanotyrannus* skull suggested that experts really had been finding juvenile *T. rex* for decades but had not recognized them.

A tyrannosaur fossil nicknamed "Jane" was proposed to be an example of "*Nanotyrannus*," until paleontologists realized that it was a juvenile *T. rex*—underscoring how much *T. rex* changed as they grew.

No dinosaur garners as much attention or controversy as *T. rex*, of course, and so multiple paleontologists became fixated on whether the famous Hell Creek Formation of the western United States truly held a small and a large tyrannosaur species or just one species that changed dramatically as it aged. Small tyrannosaur fossils had long, shallow snouts, a greater count of blade-like teeth, and lacked the bone-crushing skull architecture of adult *T. rex*, and so all these features would have to change as *T. rex* approached adulthood. Strange as it may seem, this is precisely what happened. As experts have found additional fossils of young tyrannosaurs, such as a specimen, nicknamed "Jane," that was fourteen years old when they died, they've assembled a continuum of *T. rex* forms that shows how the dinosaur went from svelte, lanky juveniles to burly adults capable of deconstructing entire *Triceratops* carcasses.

T. rex was an oddball even compared to other tyrannosaurs. Related species such as *Daspletosaurus* and *Gorgosaurus* were relatively small for the first few years of their lives, sharing the big-eyed, shallow-snouted look with the skull Gilmore described. At around ten years of age, these other tyrannosaurs began to grow more rapidly, reaching their adult forms at about twenty. Those shallow-jawed, leggy juveniles transformed into bulkier adults with thicker teeth and deeper skulls that could have supported massive muscle attachments for breaking down bones and tearing muscle from carcasses. *T. rex* took the same pattern to an extreme. Young *T. rex* stayed small and somewhat awkward until about ten years of age, but their growth spurt led these dinosaurs to rapidly put on thousands of pounds. A fifteen-year-old *T. rex* similar to

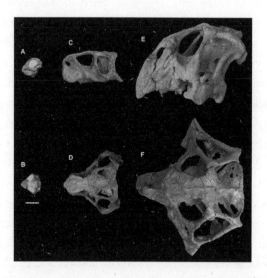

Baby dinosaurs often looked very different from adults, which has sometimes led to them being named as different species. This *Psittacosaurus* growth series illustrates how much the horned dinosaur changed as it grew up.

"Jane" would have weighed about 3,300 pounds (1,500 kg). By twenty years of age, that same *T. rex* would weigh more than 11,000 pounds (5,000 kg) and have a thick frame with a deep skull full of teeth adapted to puncturing more than slicing. Within a five-year span, *T. rex* grew so fast and changed so much that they seemed like different animals. Different as the skull might have initially seemed, the one Gilmore described was from a young *T. rex* that perished in its adolescence.

The fact that juvenile and adult *T. rex* were so different that they were confused for different species reveals the unique way many dinosaur ecosystems worked. When we look at ecosystems with an array of carnivorous species today, such as the grasslands of eastern Africa where lions and spotted hyenas roam, there seems to be an array of meat eaters at small, medium, and large sizes. Such predators have different hunting styles and prey preferences, allowing them to coexist with each other. Paleontologists have found a different pattern among carnivorous dinosaurs.

Even the largest of the predatory dinosaurs, such as *T. rex* itself, hatched out of relatively small eggs. The dinosaurs grew fast but still took years to reach their adult size, going from chicken-size to 9-ton (8,000 kg) giants in about twenty years. And while adult *T. rex* could both hunt and dismantle multi-ton carcasses of the duck-bill *Edmontosaurus* and the horned *Triceratops*, juveniles of the same species did not have the strength to take down or break apart such hefty prey. Instead, they fed upon the smaller and medium-size dinosaurs of the same habitats, slicing off muscle and swallowing some body parts whole rather than shattering skeletons. The adolescent *T. rex*, in other words, filled the distinct ecological role of the medium-size predators that we would expect to find, preventing other dinosaur species from taking it up. It's one reason that the Hell Creek Formation, in which *T. rex* is found, sometimes seems to have lower dinosaur diversity than other dinosaur habitats, a smaller number of species taking up more space on the landscape. The split may even have helped young *T. rex* stay safe from the adults. Bitten bones indicate that *T. rex* had no qualms about eating their own—at some point, meat is meat—and so favoring different prey would have reduced the chance that a "Jane"-size *T. rex* would have run into confrontations with adults.

Hatchling and juvenile dinosaurs are still rare in the fossil record. Often paleontologists know species only from the sub-adult animals that were able to survive the stresses of their first years but did not reach their full size, succumbing to drought, flooding, disease, predation, and all the other familiar dangers animals face today. But there has been the slow realization that dinosaurs did not maintain the same form throughout their lives. Dinosaur babies were big-eyed, awkward reptiles that would have been as adorable as puppies, gradually growing their distinctive crests, horns, and even bites with time. The transformation was a large part of dinosaurian success, an ongoing transformation that allowed young dinosaurs a chance to survive alongside their imposing parents.

CHAPTER 4

Hot-Running Dinosaurs

Reptiles have long suffered from a bit of an image problem. Despite the fascinating array of shapes they've evolved into, their colorful scales, and unique behaviors, the lizards, crocodiles, turtles, and other reptiles of our planet are often lumped together as "cold-blooded." The phrase often evokes a lack of feeling, associated with sluggish locomotion and low intelligence. Even among humans, some still parrot the outdated concept of the "reptile brain" that was invented to explain our baser, more reactionary urges, as if our ancestors had to overcome reptilian baggage to become human. The phrase serves only to underscore how low our opinion of reptiles is, as does the term "crocodile tears" and referring to pathologically harmful actions as done in "cold blood." It's a mammalian conceit, one that assumes our near-constant warm body temperatures make us superior; after all, an "Age of Mammals" blossomed after all those ancient, terrible reptiles were out of the way. We have long seen ourselves as improved upon and above the reptiles, as if nature formed a hierarchy.

Dinosaurs, of course, were drawn into these stereotypes, too. Throughout much of the twentieth century, dinosaurs such as *Stegosaurus* and *Tyrannosaurus* were depicted as little more than giant lizards, slow-moving and dim-witted creatures concerned only with feeding and resting. The most iconic depictions of the Jurassic sauropod *Apatosaurus*, after all, are of the dinosaur lazily munching on water plants while soaking in fetid swamps. What else could a giant dinosaur do? The biggest dinosaurs were so enormous, paleontologists calculated, that sauropods such as *Apatosaurus* and *Brachiosaurus* probably had to sun themselves for hours each day in order to heat up the muscles and organs held inside their bodies. It seemed as if dinosaurs lived in a slow-motion world, tail-dragging predators lumbering after sluggish herbivores that could escape the gnashing teeth only by running into the water. It wasn't only scientific papers, museums, and popular science books that depicted dinosaurs this way. Before the pivotal year of 1993, when the film adaptation of *Jurassic Park* debuted, dinosaurs were frequently tapped as vicious, waddling movie monsters, awkwardly jerking around the screen through stop-motion effects in movies like *The Valley of Gwangi* or even depicted as living reptiles with rubber horns and sails cruelly glued on in pulp cinema like the 1960 remake of *The Lost World*. It seemed that everyone identified dinosaurs as little more than monstrous versions of alligators and monitor lizards.

But the now-discarded image of dinosaurs as cold-blooded dullards had more to do with scientific ignorance than any specific discovery. It took time for paleontologists to meet dinosaurs on their own terms and realize that there was more to them than their strange shapes. In a sense, paleontologists had to rediscover some of the field's early hunches about dinosaurs and go back to the bones for new evidence.

During the time *Megalosaurus* and *Iguanodon* were first described in the 1820s, paleontologists were trying to understand the smattering of bones and teeth that had been found in terms of reptilian shapes. *Iguanodon*, after all, means "iguana tooth" and was originally envisioned as a supersize version of the modern lizard. But that early image didn't stick for long. Even the fragmentary fossils that had been found by 1842—when the word "dinosaur" was coined—suggested that they were different from any familiar reptile. When British paleontologist Richard Owen advised on the life-size dinosaur models that would debut at the Crystal Palace Exhibition in 1852, he directed sculptor Benjamin Waterhouse Hawkins to depict them with their legs held column-like beneath their bodies, more like rhinos than reptiles. The fossil leg bones of dinosaurs demanded such a posture, leading Owen to believe that dinosaurs were effectively reptiles evolving into mammal-like shapes. Especially during a time when evolution was often thought of as a matter of improvement or progress, dinosaurs seemed like creatures that had somehow risen above their sprawling and crawling relatives. Discoveries of dinosaurs with short front and long hind limbs like *Hadrosaurus* and the predatory *Dryptosaurus* in New Jersey, as well, hinted that dinosaurs were dynamic, active animals that were much more like birds than other reptiles. The famous 1897 painting *Leaping Laelaps* by Charles R. Knight, depicting one carnivorous dinosaur midair about to pounce on another, embodied the burgeoning idea that dinosaurs were unique creatures that were more agile and behaviorally complex than what zoologists expected of other reptiles. Even though paleontologists could not exactly explain it, dinosaurs simply looked like they had to be active creatures with more in common with warm-blooded animals than their reptilian relations.

This 1897 painting by Charles R. Knight depicts the outdated idea that large sauropods were unable to walk on land and spent most of their time wallowing in swamps.

By the beginning of the twentieth century, however, paleontologists had almost entirely abandoned the idea that dinosaurs were anything more than reptiles that grew too large, too laden down with strange ornamentation and headgear to be capable of the warm-blooded feats earlier paleontologists had hypothesized. Instead of investigating why dinosaurs were so unusual, paleontologists took a more conservative approach that assumed dinosaurs lived by the same rules as extant reptiles. So began a retrograde interpretation of dinosaurs that would last for decades, underwritten by the fact that paleontologists were working from myths and misunderstandings about living reptiles. The extended amount of time it took for zoologists to recognize parental care in alligators and crocodiles is an instructive example. It's well-known that crocodylians are doting parents, standing watch over their nests as their eggs incubate and helping dig their offspring out when they hear their little ones begin to gulp and chirp from inside their shells. Some, such as American alligators, even delicately carry their young around in their mouths for days after, guarding their babies from predators as the little snappers learn to swim and hunt on their own. Curious people had been documenting such caring behavior since 1792, at least, but parental care in crocodilians was treated as a myth until 1971, despite the fact it could be readily seen by anyone who observed the reptiles carefully. Experts didn't think reptiles would be behaviorally complex or caring parents, and so the obvious went unrecognized. As believed by zoologists and paleontologists in the twentieth century, reptiles were largely deemed inferior to mammals, and so dinosaurs became subject to the same stereotypes. Especially during a time when paleontologists were looking for good evidence of transcendent evolutionary change in the fossil record, dinosaurs were considered to be too strange and too rare to demonstrate anything interesting about life's broader story, good for drawing crowds to museums but little else. And as the public began to connect the idea of "going the way of the dinosaur" to big businesses that grew too large to adapt to changing markets, the mythology of the dull and drab dinosaur crawling through the Jurassic stuck.

Even so, a few dinosaurs didn't match the broader narrative of dinosaurs as giant, slow lizards. In 1903, paleontologists working at a site called Bone Cabin Quarry in Wyoming turned up a gorgeous skeleton of a small carnivorous dinosaur. About the size of a turkey, the dinosaur had clawed hands, many sharp teeth, and a light build. The Jurassic creature was named *Ornitholestes*—or "bird robber"—and was often depicted as trying to catch early birds like the feathery *Archaeopteryx*. The dinosaur simply looked agile, its tail a useful counterbalance rather than a limp noodle dragging behind. Instead of inspiring paleontologists to second-guess their broader assumptions, however, *Ornitholestes* seemed to be a rare exception to what paleontologists generally saw as reptiles too big and strange to survive. The whole of the Mesozoic era was a cold-blooded interlude that delayed the rise of the mammals for too long.

Ornitholestes was initially envisioned as a warm-blooded "bird-robber" that hunted smaller animals.

Theory must always contend with fact, of course, and dinosaur skeletons simply did not comply with the tail-dragging slowpoke image. Mounted dinosaur skeletons like herbivorous *Iguanodon* found in a Belgian coal mine, placed on display at the Royal Belgian Institute of Natural Sciences, had to have their tails broken to make the droopy-tail posture work. (Like many ornithischian dinosaurs, *Iguanodon* had stiff tails that were further supported by stiffened tendons and could not be dragged like lizards' tails.) Mounts of horned dinosaurs such as *Triceratops*, too, had their arm bones effectively dislocated in order to make a sprawling posture possible for the reconstructed skeletons, and paleontologists realized that if dinosaurs like *Brachiosaurus* bathed and required the water to hold up their bulk, then the Jurassic must have been dotted with an abundance of extra-deep lakes that no one had ever found evidence of. Dinosaurs were literally being bent out of shape to fit a particular image. Then, just as paleontologists were beginning to grow curious about dinosaurs again, an assemblage of bones in Montana provided a spark that would help reignite visions like Knight's *Leaping Laelaps*. The rediscovery of a mysterious dinosaur in 1964 helped revive the idea that dinosaurs were active animals with elevated body temperatures and complex behaviors.

Late in the summer of 1964, paleontologist John Ostrom and colleagues were searching for fossils around the town of Bridger, Montana, when the team found a bone bed brimming with skeletal parts. The team collected hundreds of bones over the next two years, most coming from a small carnivorous dinosaur and others from a medium-size beaked herbivore. Upon examination, the carnivorous dinosaur bones corresponded to fossils found in Montana three decades earlier, a creature that had been informally called "Daptosaurus" but never fully described. The dinosaur wasn't just another *Ornitholestes* or other small carnivore. The predator had an impressive claw on each foot that was held off the ground on an hyperextendable second toe, and the tail vertebrae had interlocking struts that required the appendage to be held aloft behind the body for balance. Despite lacking a single, complete skeleton, the historic fossils and the pieces Ostrom's team collected suggested a 10-foot-long (3 m) predator that must have been an agile hunter. The fact so many bones of the carnivore were found with those of an herbivore, too, hinted at some form of social behavior that hadn't been considered for dinosaurs before. Perhaps this small carnivore was a pack hunter of some kind, working in groups to bring down prey too large for any individual alone. Ostrom finally unveiled the dinosaur in 1969, naming it *Deinonychus*—the "terrible claw."

The discovery of *Deinonychus* required that paleontologists revise decades of scientific assumptions about dinosaur lives. The find was also announced just as the broader discipline of paleontology was about to be reshaped by personal computers. Scientists were able to create databases of fossil occurrences, create two-dimensional models to compare fossil shapes, and more, giving experts powerful new tools to investigate the history of life. The historic emphasis on simply finding and describing fossils gave way to a "paleobiological revolution" in which paleontologists began to explore how prehistoric organisms actually lived and what they might be able to tell us about the big picture of life on Earth. Previous ideas of orderly progression from the ancient to the modern were replaced by visions of unusual ecosystems and mass extinctions, devastating events that forever changed evolutionary history—including a disaster that wiped out non-avian dinosaurs seemingly overnight. Both thanks to new finds and new scientific tools, paleontologists were becoming curious about dinosaurs again, conducting research on aspects of dinosaur history that were previously thought to be beyond the reach of scientific inquiry.

Dinosaur paleontologists began to question the received wisdom of their field. There was no actual evidence that dinosaurs like giant *Apatosaurus* wallowed in mucky swamps, for example. It was just a convenient image of what a big, cold-blooded animal might do based upon a guess. Dinosaur

The discovery of *Deinonychus* spurred paleontologist John Ostrom to reconsider how active and warm-blooded dinosaurs may have been, helping to catalyze the "Dinosaur Renaissance."

tracks rarely showed any signs of tail drags, as well, corroborating skeletal hunches that they held their tails aloft and used them as counterbalances as would be expected of an active and behaviorally complex animal. The relative abundance of dinosaur herbivores compared with carnivores, too, made more sense considering the dynamics between mammal predators and prey. Carnivorous mammals like jaguars are rare compared to their prey, in part, because they maintain high, constant body temperatures and must eat relatively often to meet their energetic needs. If there are too many jaguars, then there soon won't be enough capybara and armadillos to eat. The number of jaguars must necessarily be only a fraction of their prey's population size. Cold-blooded animals like caimans, however, are not constrained because their body temperature is regulated by the environment and their energetic needs are less. The reptiles don't need to eat as often, and so a greater number can exist in a population than if they were warm-blooded. The proportion of dinosaur carnivores like *Allosaurus* to sauropod herbivores like *Camarasaurus* in the same environments better matched the warm-blooded mammal population ratios rather than those of cold-blooded reptiles. The rarity of dinosaur carnivores made sense if they were warm-blooded, but would be a mystery if they were cold-blooded. Even the internal structure of dinosaur bone underscored that dinosaurs were not just big lizards or crocodiles. Experts began cutting into dinosaur bones to get some sense of how quickly the reptiles grew, finding fast-growing bone tissue rather than crocodile-like growth patterns previously assumed. Experts were beginning to uncover nesting grounds, bone beds of drowned herds, and other evidence that dinosaurs were more behaviorally complex and flexible than had been previously explored, driving a period of intense debate and rapid discovery called the Dinosaur Renaissance.

The question of dinosaur physiology was at the center of the Dinosaur Renaissance. Experts were amassing plenty of circumstantial evidence that dinosaurs were not just lazing through the Mesozoic period, but the assessment of whether dinosaurs maintained their internal body temperatures, fluctuated with the environment, or something else was a challenging task given experts were more than sixty-six million years too late to take dinosaur temperatures directly.

The terms warm-blooded and cold-blooded are only the roughest shorthand for the complicated relationship between body temperature and how a living thing maintains those temperatures. A lizard basking in the sun will have a high body temperature and technically be hot-blooded, while some mammals can lower their body temperatures while in hibernation or in torpor, a temporary hibernation to conserve energy. What the opposing terms are really referring to is the way in which body heat is regulated by a variety of different mechanisms and strategies.

In broad strokes, a warm-blooded dinosaur would generate their own body heat internally—endothermy—and keep that body temperature relatively constant: homeothermy. (We are endothermic homeotherms, for example, hence the oft-quoted resting body temperature of 98.6°F/37°C.) A cold-blooded dinosaur, by contrast, would be largely dependent on the surrounding habitat for their warmth, which is called ectothermy, and so their body temperature would fluctuate with their surroundings. Such a dinosaur would have to warm in the sun or move into the shade or water when too hot, the animal's behavior through the day and in each moment affecting its internal temperature. Then again, there are creatures like some sharks that maintain body temperatures above the surrounding environment even as those temperatures change, sometimes called mesothermy. And because of the relationship between surface area and internal volume, some creatures are gigantotherms—not technically warm-blooded, but big enough that their bodies are good at retaining heat, so they have an alternate route to maintain elevated body temperatures. The debate about hot-blooded dinosaurs wasn't so much about whether dinosaurs were capable of behaving more like mammals or birds, but what kind of physiological profile they possessed.

In 1980, researchers gathered at a symposium in Washington, DC, to discuss the possibilities, soon after published in a book called *A Cold Look at Hot-Blooded Dinosaurs*. The conference did not come to a concise conclusion. It would have been impossible. Despite the fact that we often think of dinosaurs as a singular group of animals with many shared traits, they were as diverse and disparate from each other as mammals. The physiologies of a bowhead whale, a little brown bat, a kangaroo, a duck-billed platypus, and a human are not going to be identical despite the fact all are mammals. Neither would dinosaurs all follow a single physiological profile. Paleontologists at the conference could affirm that the old image of slow, long-lived dinosaurs reliant on an endless Mesozoic summer should be discarded, but the symposium concluded with more questions than answers.

So far, the precise details of dinosaur physiology remain difficult to approach. A lack of living non-avian dinosaurs certainly complicates our ability to understand how their bodies functioned. Still, paleontologists have uncovered an incredible amount of evidence that dinosaurs were active, fast-growing animals that likely had elevated body temperatures. How the reptiles accomplished this and how they differed from each other has been more difficult to determine, paleontologists looking to everything from the biochemistry of dinosaur eggshell to the geographic distribution of dinosaur species to

The large tyrannosaur *Yutyrannus* had a thick coating of feathers and lived during a time when its region of China was relatively cold, hinting that it was a warm-blooded animal that needed feathers for insulation.

investigate their physiology. One 2022 study by paleontologist Alfio Alessandro Chiarenza and colleagues, for example, compared prehistoric climate models with the geographic spread of long-necked sauropod dinosaurs like *Patagotitan* and *Brachiosaurus*. The experts found that sauropod dinosaurs tended to live in lower latitudes that were warmer, areas that were tropical and subtropical during the Jurassic and Cretaceous, whereas other dinosaur groups expanded from pole to pole. The spread might hint that sauropod dinosaurs had a greater difficulty maintaining their body temperatures in colder habitats, perhaps meaning that they were more ectothermic and relied on their considerable size to maintain body heat. Another recent study by paleontologist Amzad Laskar and colleagues seems to back up this hypothesis. Geochemical markers in dinosaur eggshell recoded the parent's body temperature when the eggs formed, which could then be compared to estimates of prehistoric climate. The researchers found that small theropod dinosaurs had elevated body temperatures compared to their environment, but that eggs attributed to big sauropod dinosaurs reflected body temperatures closer to that of their habitat. The correspondence could mean that sauropods were ectotherms that tried to retain their body heat, or perhaps they had some as-yet-unknown adaptation to vary their body temperature as they would have a narrower window of safe body temperatures—overheating as a large animal that can't cool quickly can be deadly. But what would seem to make sense isn't always true when it comes to living things.

Other research teams using different methods have resolved different patterns for dinosaur physiology. A study by paleontologist Jasmina Wiemann and colleagues also published in 2022 compared the biochemical clues related to metabolism in non-avian dinosaur bones with those of living birds and reptiles. The researchers concluded that the ancestors of dinosaurs and the flying pterosaurs were already endothermic, with sauropod and theropod dinosaurs retaining the ancestral, endothermic state. Ornithischian dinosaurs like *Ankylosaurus* and *Parasaurolophus* showed signs of becoming more like ectotherms during their history and, as Wiemann and colleagues suggested, were more likely to behaviorally manage their body temperatures.

Paleontologists have discovered entire ecosystems in Alaska that would have been within the prehistoric Arctic Circle during the Cretaceous. Baby dinosaur fossils, in particular, indicate that dinosaurs lived through the dark, cold winters in the high latitudes.

Some of the best evidence that dinosaurs were warm-blooded doesn't come from the dinosaur bones themselves, but where they were found. Alaska's Prince Creek Formation represents what's left of a sodden coastal plain that was located within the ancient Arctic Circle about seventy million years ago. The rock layers are full of dinosaurs. The horned dinosaur *Pachyrhinosaurus*, the small tyrannosaur *Nanuqsaurus*, and many more—some as-yet-unnamed—have been found within the formation, along with various fossil mammals and plants. What experts haven't found, however, are reptiles like lizards, crocodiles, and turtles that are commonly found in other dinosaur-bearing layers around the world. The presence of dinosaurs, but an absence of ectothermic reptiles, is a potent clue.

As *Pachyrhinosaurus* was moving in herds along the muddy coast of ancient Alaska, the habitat experienced the same seasonal light shifts of northern Alaska today. Summer months of near-endless light shifted into Arctic winters where temperatures plummeted and sunlight was absent for months. Ectothermic animals struggle in such habitats as they have no way to warm up or stave off the cold without near-daily sun exposure. Endothermic animals can maintain their body temperatures through the chill, however, which would explain the presence of warm-blooded dinosaurs and ancient mammals but not ectothermic reptiles. And while paleontologists once thought that Alaska's dinosaurs would have migrated according to the seasons, the discovery of embryonic dinosaur bones within the Arctic Circle indicates that dinosaurs were breeding, nesting, and growing up in this extreme habitat. Dinosaurs ran hot enough to withstand the depths of the northern cold year after year, the best direct evidence that dinosaurs evolved their own forms of warm-bloodedness and were capable of living everywhere from sweltering, dune-covered deserts to dark Arctic forests.

Even as paleontologists have begun to detect and define some subtle parts of dinosaur lives that were thought to be beyond scientific reach, such as dinosaur color, dinosaur physiology remains an area where more questions remain than answers uncovered. The research is still coming up with complicated and sometimes conflicting evidence, as one would expect as scientists attempt to study long-extinct dinosaurs as living systems. Somehow or other, most dinosaurs were warm inside. The mechanics of how will be difficult to tease out, but even in its short history, paleontology has managed to explore an entire array of dinosaurian particulars that were at one time believed to be scientifically undiscoverable. Perhaps one day someone will work out how to reliably take dinosaur temperatures and a clearer understanding of hot-blooded dinosaurs will emerge.

CHAPTER 5

The Largest Creatures to Walk the Earth

Of all superlative dinosaur traits, few are as celebrated as size. Every few years, headlines appear about a new contender for the largest dinosaur of all time, leading to debates among both paleontologists and the public alike about who the largest was. Even putting the title holders aside, museum halls have long banked upon the imposing stature of giant dinosaurs like *Apatosaurus*, *Triceratops*, *Giganotosaurus*, and many more, creating the impression that everything was bigger during the Mesozoic era. The truth is that the real mark of dinosaur success is the vast range of sizes the animals evolved into—from sparrow-size to titans longer than a blue whale—but it can't be denied that many, *many* dinosaur species were much larger than any terrestrial animals we know of today. The unique natural history of dinosaurs helps to explain why the reptiles were able to reach proportions that have never been matched by any land-dwelling mammal and how so many of the great saurians were able to live side by side.

One of the best places in the world to understand the sheer scale of the Mesozoic straddles the Utah-Colorado border. Enclosed in the tan sandstone wall of the quarry building of Dinosaur National Monument are the bones and body parts

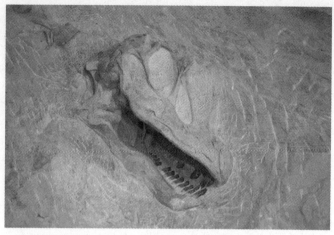

A *Camarasaurus* skull preserved within the quarry wall of what is now Dinosaur National Monument, Utah.

of various giant dinosaurs that lived in the region around 150 million years ago. Sauropod dinosaurs are the most common among the Jurassic remains. The spoon-toothed skull of a *Camarasaurus* grins from the tilted wall above the limb bones of *Apatosaurus*, a stone's throw from a tail segment from *Diplodocus* and *Barosaurus* vertebrae. That's to say nothing of the bones of the plate-backed *Stegosaurus*, carnivorous theropod *Allosaurus*, and other creatures exposed within the quarry wall. Even long dead and dismembered, it's not difficult to bring those bones back together and imagine these enormous creatures wandering over a fern-covered plain dotted with conifers and ginkgo trees. By this Late Jurassic period, dinosaurs were undoubtedly living large.

The quarry that you can visit today is just a fraction of what once existed at the site, leaving us to imagine what the entire bone bed might have looked like if some of the larger and more complete skeletons were left in the rock. To see those, you have to travel over 1,700 miles (2,735 km) to the east to see them on display at the Carnegie Museum of Natural History in Pittsburgh, Pennsylvania, reconstructed and revived in lively poses among a facsimile of their Jurassic home. It was paleontologist Earl Douglass, under directions from the museum to find big dinosaurs for their halls, who first spotted a string of *Apatosaurus* vertebrae jutting out of the site's rock in 1909. He began digging and kept going, and going, not only excavating skeletons to send back by rail to Pittsburgh, but campaigning for the fossil site to be preserved as a national park. He was granted his wish in 1915, and ever since visitors from all over the world have come to the cottonwood-dotted desert to stand in the presence of gargantuan reptiles far larger than any terrestrial creature alive today.

Apatosaurus, *Camarasaurus*, and *Diplodocus* were not the largest dinosaurs to have ever lived. That title is contested among Cretaceous dinosaurs such as *Patagotitan* and *Argentinosaurus*, sauropods that exceeded 100 feet (30 m) in length and weighed more than 70 tons (63,500 kg). And even in the preceding Late Jurassic, there were larger dinosaurs than the classic trio from the Late Jurassic rocks of western North America. The slender *Barosaurus* and *Supersaurus*, rarer species found in the same stacks of Late Jurassic rocks, were longer, and the thick-armed *Brachiosaurus* was heavier, itself once the titleholder for largest dinosaur of all time. Nevertheless, running a "my dinosaur is bigger than yours" contest is irrelevant to the larger point that this constellation of Jurassic species makes. Sauropod dinosaurs spun off gigantic species time and again, sometimes with several gargantuan species living side by side as they did in the famed Morrison Formation habitats. How such dinosaurs evolved to such stupendous sizes has entranced paleontologists for decades.

Diplodocus was one of the largest sauropods of the Late Jurassic. The largest *Diplodocus* species, *D. hallorum*, grew to more than 100 feet (30 m) in length.

The fact that nothing as large as an *Apatosaurus* walks the Earth today certainly molds our fascination. Compared to the Late Jurassic days now locked into Morrison Formation stone, we live in a time pitifully devoid of megafauna. And accounting for extinctions of large animals that have been transpiring since the close of the Ice Age, even the largest mammoths could have stood in the shade of the biggest sauropods. Only whales have approached and exceeded the biggest sauropods in size, but even then they can be only a rough analogy given the different evolutionary pathways opened by water. On land, nothing has ever grown larger than the sauropods. Not even close. Even the largest terrestrial mammals, such as the roughly thirty-million-year-old rhino *Paraceratherium*, only got to be about 15 feet (4.5 m) tall at the shoulder and weigh about 15 tons (13,600 kg), puny compared to many sauropod dinosaurs. Something must have been different about the life and times of the Morrison Formation sauropods and other giant dinosaurs. The answer doesn't reside within a single cause, but instead a combination of influences that opened up possibilities for dinosaurs that would be impossible for mammals to duplicate.

Part of the secret to supreme dinosaur size was the very thing that allowed the "Age of Reptiles" to kick off in the first place—eggs. So far as we know, every single dinosaur to ever walk the Earth started life by hatching out of an egg. Even the largest dinosaurs of all pushed their way out of eggs between the size of a large grapefruit and a soccer ball, packing on hundreds and hundreds of pounds each year in a race to get too big to be munched by predators. Paleontologists have found the nesting grounds of these giants in Patagonia, India, and elsewhere, some including rare embryos that have revealed just how much sauropods changed from the day they hatched out of their eggs to adulthood. The round eggs huddled together in these nests opened evolutionary possibilities that allowed dinosaurs to embody a range of sizes from tiny to immense.

To understand why sauropod eggs were so important to their size, compare the life histories of the big dinosaurs to those of big mammals such as elephants, giraffes, and rhinos. Despite their different evolutionary histories, large mammals around us today typically have a single offspring that gestates within its mother for a very long time. Mother elephants are pregnant with their babies for up to two years. That is an incredible physical investment into a single offspring. Even then, the babies require a great deal of attention, socialization, and, especially, milk, a protein-rich liquid that is incredibly taxing for mother mammals to produce. For mammals to become comparable in size to sauropods, then, they'd likely have to gestate their babies for even longer periods, giving birth to bigger babies that required extended parental care, and then the process would have to repeat over and over. The longer a large mammal's pregnancy goes on, however, the more chances there are for something to go awry—either to the mother herself or the developing embryo. Carrying offspring internally, in other words, requires a specific set of biological changes that actually limits how large land-dwelling mammals can become. Bigger beasts would require longer gestation periods and extended windows of parental care, making such big species more susceptible to population crashes and extinction. Dinosaurs experienced no such limits. In fact, paleontologists estimate that truly giant sauropod dinosaur lineages evolved at least thirty-six times during a span of over 180 million years. For sauropod dinosaurs, it was apparently easy to get big.

But eggs are only part of the story. The entire sauropod body is a testament to the unusual changes required to not only grow to huge sizes but maintain bodies that weighed tens of tons. Another one of their secrets, shared with the carnivorous theropods who often fed on them, was air sacs.

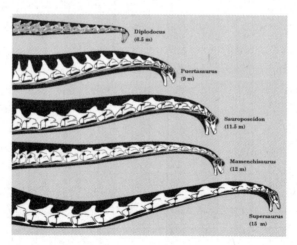

Sauropod necks were extreme even when compared to other long-necked creatures. Air sacs in their neck vertebrate helped keep the bones relatively light and opened possibilities for the dinosaurs to grow longer than most other animals.

One of the key features that unites saurischian dinosaurs is the presence of air sacs stemming from their respiratory system. You can see the evidence among the indentations and pockets within their bones. The neck vertebrae of sauropod dinosaurs, especially, are often so riddled with indentations and hollows that they are the most distinctive parts of the skeleton for telling one species apart from another. In technical terms, paleontologists call this pneumaticity—the indentation and even invasion of air sacs connected to the respiratory system in the bone. Consider the neck vertebrae of that classic Jurassic sauropod, *Apatosaurus*, for example. The dinosaur's neck bones look incredibly complex, with thin struts and wings of bone around the cylindrical centrum that forms a vertebra. When paleontologists look inside the bones, they find hollows and pockets that once housed air spaces. We know this because the skeletons of birds show the same kind of air pocket extension, and these extra air spaces provide an array of advantages that made supersize dinosaurs possible.

Air sacs are part of what drove sauropods, like their bird relatives, to breathe in a different way than we do. We mammals employ tidal breathing, a big inhale to bring more oxygen-laden air in and an exhale to blow the relatively oxygen-depleted air out. But in sauropods, birds, and other dinosaurs, incoming air moves in a unidirectional fashion, new air pushing the older air through the series of air passages inside the animal. The greater surface area of the air sacs and the one-way passage allows the animals to draw more oxygen from the air, making their breathing more efficient. It's part of what enables birds to fly at high altitudes, and it would have allowed sauropods to draw in the oxygen they needed for their big bodies. The same airway system also would have helped sauropods keep cool. The reptiles didn't sweat, but air moving throughout their bodies could have picked up any extra heat, sweeping it out of the body as it cycled through. Considering how bigger and bigger animals have a difficult time with dumping excess body heat because of their increased internal volume compared to the surface area of their skin, such a cooling system was a lifesaver. And, of course, effectively replacing bone with air spaces kept sauropod skeletons relatively light without sacrificing strength. While it might seem a little silly to call a 100-foot-long (30 m), 45-ton (40,000 kg) animal like *Supersaurus* "light," the fact is that a mammal of the same skeletal stature—lacking the additional air sacs—would be much heavier, partly in requiring more muscle to move the heavier bones.

But one of the most potent secrets to supreme sauropod size has been hiding in plain sight all along. The basic layout of the sauropod body—a tiny head with simple teeth, a long neck, and a big body counterbalanced by a long tail—is part of what allowed sauropods to grow so big.

Once again, big mammals offer a helpful comparison. Large herbivorous mammals of the past and present—whether we're talking about giant rhinos like *Paraceratherium* or today's giraffes—have complex teeth and powerful jaw muscles that allow the plant eaters to chew vegetation to pulp. Chewing requires teeth that come in contact with each other, muscles to carry out the repetitive motion, and thick neck muscles to hold those heavy heads full of teeth and muscle up. Each of these parts go together, meaning herbivorous mammals can process their food extensively before swallowing. Sauropods did something else. A dinosaur like *Brachiosaurus* nipped and cropped vegetation from their habitat, swallowing plants whole. All the breakdown happened inside their bodies, between their stomach and intestinal tract. Long necks helped these dinosaurs stand in one place and pluck off whatever food they could reach, effectively stuffing themselves with unchewed vegetation that their bodies would then break down. Small, light heads allowed necks to grow long, although all the bone, muscle, and other tissues of the neck had to be counterbalanced by tails that were just as long, if not longer. Sauropods evolved to be eating machines, maximizing their reach and intake instead of stopping to chew their food. Young sauropods had more stringent nutrition needs to grow fast, packing on as many as 1,000 pounds (450 kg) in a year, but the general pattern was the same. The largest dinosaurs evolved to bulk feed to an incredible degree, their bodies adapted to reaching green food near and far, high and low.

Eggs, air sacs, and tiny heads on long necks help explain only what *allowed* sauropods to grow much larger than mammals, however. What we're missing is the reason why sauropod dinosaurs became giants dozens of times over. A definite answer is difficult to obtain, especially because sauropods repeatedly evolved into giants. Even within the Morrison Formation ecosystems that hosted several different species of giants, the array of dinosaurs represents at least three major sauropod groups that all evolved into huge sizes and perhaps did so for different reasons. The ecosystems, plants, and other dinosaurs in each setup for gigantism might have differed from one scenario to the next, meaning that there was no single route for evolution to fashion a giant reptile. Nevertheless, there's one consideration that most if not all sauropods had to contend with—growing large enough, fast enough, to ward off predators.

All sauropods started off life small, around the size of a house cat. Many lived in the same habitats as a range of

THE LARGEST CREATURES TO WALK THE EARTH   93

carnivorous dinosaurs, from chicken-size hunters to meat-eating giants such as the knife-toothed theropod *Torvosaurus* of the Jurassic or even the Cretaceous *Tyrannosaurus* itself. Hatchling and juvenile sauropods would have been prime prey for such carnivores, too small to effectively defend themselves or make predators wary of injury. Paleontologists have speculated that the extreme rarity of hatchling and juvenile dinosaurs may be a reflection of this pattern that's also seen among modern carnivores—young, small, naïve herbivores are easy prey. The only sure route for sauropod dinosaurs to avoid becoming dinner was to grow big, and fast. A yearling *Apatosaurus* would have been a snack to a predator like *Allosaurus*, but an adult animal would have been large enough to break bones or even stomp a brazen carnivore. Size offered safety.

The constant back-and-forth between predator and prey molded ancient ecosystems for millions of years. Sauropod nesting grounds would flood ancient habitats with seemingly innumerable offspring, much like baby sea turtles emerging in droves on beaches today. Predators would have no doubt taken advantage of the sudden glut of easy-to-catch, inexperienced prey. But sauropods who were able to survive their first year were much more likely to survive to adulthood, when they could more effectively ward off all but the most desperate of carnivores. A big *Diplodocus* or *Camarasaurus* might live many more years relatively unbothered, and when they eventually perished, they left tons and tons of meat for those same carnivores to scavenge, forming the backbone of diverse dinosaur ecosystems the world over.

The repeated evolution of the giants also offers us another insight into the biology of dinosaurs. Even though paleontologists seem to announce a new "biggest dinosaur" every

The title for longest dinosaur of all time is disputed, but *Mamenchisaurus* had one of the longest necks. An 85-foot-long (26 m) *Mamenchisaurus* would have had a neck over 40 feet (12 m) long.

few years or so, the fact of the matter is that the longest, heaviest sauropods are all very close to each other in size. The Jurassic *Supersaurus* and *Diplodocus hallorum* (once known as *Seismosaurus*) are almost identical in size, just as the later *Argentinosaurus*, *Alamosaurus*, *Puertasaurus*, and *Patagotitan* were heavier but still close to the same overall size. Contenders for even longer, heavier dinosaurs—such as *Bruhathkayosaurus* from India, or the *Maraapunisaurus*, what was once called *Amphicoelias fragillimus*—are based on incomplete, lost material that can't accurately be assessed. It's possible that the close competition between these dinosaurs indicates that growing to be more than 100 feet (30 m) long and weighing more than 45 tons (40,000 kg) was pushing the limits of what was possible for dinosaurs. Then again, even the biggest dinosaurs we know of probably were not the largest of their own species.

Individuals of any given species vary from one to the next. Simply by walking down the street, for example, we can see that even people our own age can be much taller or shorter than us, often due to their own genetic background and how they grew up. The same is true for animals, as there is an average size many individuals cluster around in addition to outliers that can be significantly larger. The trouble with dinosaurs is that we can't simply observe populations of the reptiles to get an understanding of how they varied from each other. We have only a small fraction of individuals spread across spans of time and geography. On top of that, the truly large dinosaurs are difficult to bury. Of all the close contenders for the biggest dinosaur of all time, none are known from anything close to a complete skeleton. The size of the animals required so much sediment to be buried that their bodies undoubtedly sat exposed to scavengers and the elements before being buried, usually in jumbles rather than articulated parts. This is particularly true for the bigger dinosaurs. The upper limit of 100 feet (30 m) and more than 45 tons (40,000 kg), then, might not represent a biological ceiling but a geological one. It may be that dinosaurs surpassing such sizes are incredibly rare, perhaps preserved only as a few bones out of hundreds that comprised their skeletons. Paleontologists estimate that such rare, giant-size members of their species may have been up to 70 percent heavier than the largest known specimens, but were likely so few that it may take centuries to discover direct evidence of their existence. Any time you see a giant dinosaur in a museum hall, there is a good chance that other individuals of the same species grew even grander still. For every dinosaur we know, there may be record-breaking individuals still in the rock.

CHAPTER 6

Dinosaurs of a Feather

In the Early Cretaceous period, more than 125 million years ago, in what's now China, the body of a fuzzy little dinosaur was blanketed by sediment at the bottom of a prehistoric lake. The animal was not especially unusual or uncommon for its time. A miniature carnivore about the size of a raven, the dinosaur was covered from head to toe in simple fuzz, shaded in red and white almost like the red pandas that would evolve tens of millions of years later. There the carcass rested as it decayed, minerals borne by water gradually replacing bone and fuzz, preserving both skeleton and feathers together.

If the fossil had been discovered during the nineteenth century, and if the science of paleontology had been aware of fossil knowledge in east Asia, the fossil might have significantly altered how paleontologists reconstructed the history of life. Not only was the fossil considerably more complete than the dinosaur fossils being turned up in England and other parts of Western Europe, but it would have left no doubt that dinosaurs of the whole Mesozoic period wore coats of feathers. Paleontologists might have recognized that birds are dinosaurs much earlier, likely altering the ever-changing perceptions of dinosaur appearance and anatomy. But we'll never know how such an alternate timeline would have played out. The happenstance of fossil preservation and the history of science would make paleontologists wait almost a century and a half to confirm that birds are living dinosaurs and that feathers were an incredibly ancient dinosaur feature. The fuzzy dinosaur preserved at the bottom of the lake in ancient China would dazzle scientists, but only because experts had spent decades debating whether such a fluffy dinosaur even existed.

The first fossils of *Archaeopteryx* were found in 1859, and paleontologists are still describing new ones. This is the twelfth known *Archaeopteryx* specimen, described in 2018.

The setup for the relatively recent realization that dinosaurs were feathery creatures has its roots in the mid-nineteenth century. Naturalists and paleontologists were beginning to play with ideas about evolution during this time, but no one could really agree on how change might unfold or how life came to shift between such varied forms through time. Richard Owen, for example, toyed with evolutionary ideas but thought of them in quasi-theological terms, a simplified ancestral body plan molded through divine guidance into many different forms. Charles Darwin and Alfred Russel Wallace, of course, independently settled on the idea of evolution by natural selection, variations in living things being winnowed down by different pressures to alter species generation by generation. Darwin laid out his argument for the mechanism in 1859's *On the Origin of Species by Means of Natural Selection*, but, even though the book helped popularize the idea that evolution is a reality, many experts didn't agree that natural selection drove transformations over time. Some experts saw natural selection as a neutral or destructive force, keeping living things the same, and preferred different mechanisms with an emphasis on increasing perfection or progress through time. If Darwin and Wallace were correct, the fossil record should be full of creatures with transitional features between major groups of living things, like creatures with traits of both reptiles and birds. Darwin had none to show when he wrote *On the Origin of Species*, instead focusing on the imperfection of the fossil record to explain why the species his hypothesis suggested had not been found yet. Paleontologists, knowing the history of life better than anyone, had a difficult time reconciling this idea with their observations and tried to come up with alternate ideas that might explain why the history of life seemed to have so many gaps.

A curious skeleton soon provided just the sort of evidence the Darwin/Wallace hypothesis predicted, what paleontologist Hugh Falconer called "a strange being à la Darwin" in a private letter. The area of Solnhofen in Bavaria, Germany, was famous for its limestone, especially useful in creating printing plates. The limestone had formed about 150 million years ago when the area was an archipelago surrounded by sea and dotted with lagoons. Those calm lagoons often had a layer of oxygen-depleted water on the bottom, a dead zone where everything was calm, still, and generally free of all scavengers except for microbes. When dead sea creatures like horseshoe crabs or surface-dwellers like small dinosaurs washed out to sea by storms, floated down to these bottom waters, there were no large creatures to eat the carcasses or dismember them. Many of these remains were soon covered by sediment and began their transformation into fossils, preserving details of soft tissues as well as the harder parts. Fossil collectors knew the exquisite nature of these fossils, and in 1861 a pair of significant findings would introduce the world to the oldest known bird.

The first find, described by paleontologist Hermann von Meyer, was a single wing feather. The fossil was undeniable evidence that birds lived during the Jurassic, and so von Meyer described the fossil as *Archaeopteryx*, the "ancient wing." Not long after, from a quarry near Langenaltheim, physician Karl Häberlein acquired the bird itself. Even though the head appeared to be missing, the jumble of tan bones was surrounded by feather impressions. He sold it—a creature that had both feathers as well as fingers, hand claws, and a long bony tail—to what is now the Natural History Museum in London. The skull showed *Archaeopteryx* had teeth, as well. Birds were not only much more ancient than anyone knew, but once seemed much more reptilian.

Despite the iconic status of *Archaeopteryx* as a transitional fossil today, Darwin was reserved about the fossil's significance.

Wisely, in later editions of *On the Origin of Species* he cited the early bird as evidence that the fossil record was incompletely known and was sure to give up more unexpected surprises. His colleague Thomas Henry Huxley went further. *Archaeopteryx* indicated that birds evolved from a reptile lineage, although which one was unclear. Privately Huxley favored dinosaurs, but non-feathery dinosaurs similar to *Archaeopteryx* were found in the same rocks. How could ancestor and descendant exist at the same time? Huxley mused that the origin of birds must have happened during some earlier period that was not recorded in Earth's rocks, with both the birdlike dinosaurs and *Archaeopteryx* being evolutionary holdovers from that mysterious time. Then again, paleontologists like Othniel Charles Marsh suggested that the evidence was already obvious enough to say that birds must have evolved from among dinosaurs, but no additional fossils appeared to show how supposedly scaly reptiles could have evolved feathers.

Archaeopteryx has been an evolutionary icon since the 19th century, but only recently have paleontologists fully understood the early bird's connection to other dinosaurs.

Just as dinosaurs in general were pushed to the scientific sidelines in the early twentieth century, so was the idea that birds were dinosaurs. The proposal was brought up now and again as experts uncovered new fossils, including dinosaurs sitting in birdlike resting positions, but feathers were always the problem. Dinosaurs were supposed to be predictable reptiles, covered in scales. Without a feathery dinosaur, no one could conclusively show that birds and dinosaurs were closely connected. Paleontologists began to favor a different idea, called the thecodont hypothesis, that birds evolved from thecodonts, a fossil reptile grouping of crocodile-like creatures that lived in the Triassic, and *Archaeopteryx* became dinosaur-like only through a phenomenon of evolutionary convergence, when two lineages independently arrive at similar forms. The thecodont hypothesis lacked direct evidence just like the dinosaur origin, though, and by the 1970s paleontologists noticed connections beyond fluff and fuzz. The carnivorous dinosaur *Deinonychus*, for example, had many birdlike features in the wrists, hips, and other parts of the skeleton. Present the dinosaur with feathers on and it would look like a larger *Archaeopteryx*. Paired with new discoveries that dinosaurs were likely endothermic, behaviorally complex, and otherwise

Fossil birds with teeth, such as *Ichthyornis*, led paleontologists to suspect that birds had evolved from reptiles since the 19th century, but the relationship between birds and other dinosaurs was not clear until the 1990s.

were not merely giant lizards, the possibility that birds are dinosaurs took off once more with the Dinosaur Renaissance.

Despite growing acceptance of the idea that birds and dinosaurs shared a close relationship—a concept even mentioned several times during 1993's *Jurassic Park*—no one could find the critical fossils the hypothesis suggested should exist. Birds dated back to the Late Jurassic period, at least 145 million years ago, including complex feathers with shafts and barbs. Feathers must have evolved much earlier, perhaps many millions of years before *Archaeopteryx*. If the limestone of Bavaria was capable of preserving feathers, then there should be other such fossil sites somewhere. The fossil record should contain fossils of non-avian dinosaurs with feathers. Enter the little fluffy dinosaur preserved in the fine lakebed sediment in China, *Sinosauropteryx*.

In 1996, paleontologists Qiang Ji and Shu-an Ji thought the fossil belonged to an early bird. The Chinese-language description of the fossil mentioned feathers along the animal's back, running from the back of its head to tufts along the tail. But the fossil was just part of a whole, one of two slabs preserving the same animal. The second, better slab had not yet been described, but, as paleontologists from around the world gathered at the Society of Vertebrate Paleontology meeting in October of 1996, paleontologist Peiji Chen brought photographs to show other experts. Reporters covering the conference soon learned of the fossil and began reporting the then-unpublished find, reaching further than previous press coverage of the first fossil. In an instant, after decades of uncertainty and months after the fossil quietly received its scientific name, everyone learned that dinosaurs once had feathers.

More than 135 years passed between the discoveries of

DINOSAURS OF A FEATHER

Archaeopteryx and *Sinosauropteryx*. Paleontologists would not have to wait very much longer for additional fuzzy and feathery dinosaurs. Liaoning, China quickly became world famous as paleontologists described fossil after feathery fossil, the area boasting rocks that had formed in such a way to preserve entire skeletons and delicate features like feathers. Paleontologists had been expecting that dinosaurs closely related to birds would have fuzz and feathers, but soon experts were finding animals like the small, horned dinosaur *Psittacosaurus* with bristle-like feathers on them, too. *Psittacosaurus* is much closer to *Triceratops* on the dinosaur evolutionary tree than birds, and yet it had feathery structures on its tail. It seemed unlikely that feathers would evolve more than once, hinting that they might have existed in the last common ancestor of all dinosaurs and inherited by all the major dinosaur groups. Feathers not only preceded the origin of birds, but were a widespread and diverse dinosaur feature.

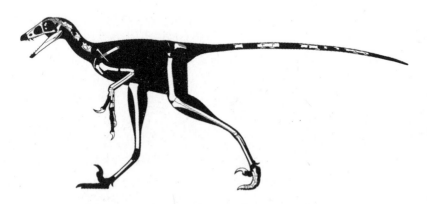

Paleontologists are still searching for the direct ancestors of early birds. Small, raptor-like dinosaurs such as *Hesperornithoides* resemble *Archaeopteryx* and may represent what direct bird ancestors looked like.

The wide spread of feathery lineages in the dinosaur family tree has complicated paleontologists' efforts to identify the dinosaur lineage birds emerged from. Various evolutionary trees favor several different possibilities. Dinosaurs like *Deinonychus* and the feathery, clawed, and winged *Anchiornis* seem to be closely related to early birds like *Archaeopteryx*, hinting that birds must have emerged from these carnivores sometime before the Late Jurassic. Then again, since 2002, paleontologists have found several tiny, feathery dinosaurs called scansoriopterygids that share a number of characteristics in common with early birds. Part of the reason for the uncertainty is that dinosaurs evolved the ability to fly at least three times—among early birds, among small predatory dinosaurs like the enigmatic, sometimes bat-winged scansoriopterygids of the Jurassic and the four-winged *Microraptor* of the Cretaceous. Instead of powered flight being a unique ability that emerged only once, various feathery dinosaurs were capable of maneuvering through the air and evolved similar skeletal traits in the process. The fact that multiple feathered dinosaur lineages evolved skeletons capable of powered flight in the middle of the Jurassic and the early part of the Cretaceous means that experts can't rely on flight feathers or characteristics related to flapping alone to work out what the direct ancestors of *Archaeopteryx* looked like. As Darwin himself observed, we still know relatively little of the fossil record. The critical fossils of direct bird ancestors have yet to be identified.

While the array of feathery fossils found so far has somewhat frustrated paleontologists' attempts to work out what dinosaur lineage birds split from, the spread of prehistoric plumage has nevertheless provided some unexpected insights into dinosaur lives. When *Archaeopteryx* was discovered, feathers were thought to be an exclusively avian feature. And even when *Sinosauropteryx* was officially described and recognized as a dinosaur, some paleontologists stressed that the plumage on the dinosaur's body were "protofeathers," or feather precursors. Then experts began finding feathers on dinosaurs more and more distantly related to birds—on tyrannosaurs, the ostrich-like theropods ornithomimosaurs, the stubby-armed and beaked alvarezsaurs, the parrot-like, winged oviraptorosaurs, and more. The discovery that some ornithischian dinosaurs, despite only a distant relationship with birds, had feathers hinted that plumage must have evolved multiple times among dinosaurs or that it was present in the earliest dinosaurs and subsequently modified or lost as the "terrible lizards" proliferated. At the same time, paleontologists were puzzled by fuzzy body coverings on some pterosaurs, which had been known since 1831. Pterosaur fuzz was thought to be fundamentally different from the feathers of dinosaurs, dubbed "pycnofibers" by experts. But in 2019, paleontologists observed that pterosaur and dinosaur fuzz are structurally the same and may have shared a common origin among their ancestors, like the miniature Triassic hopper *Kongonaphon*, living 237 million years ago. Given that dinosaurs and pterosaurs are each other's closest relatives, with an ancestor dating back to the early part of the Triassic, feathers were most likely present in their last common ancestor. The feathery inheritance might even indicate why feathers evolved in the first place, opening up various biological possibilities through the Mesozoic timeline.

Kongonaphon was not a dinosaur or pterosaur, but was close to the ancestry of both. Paleontologists also know that the earliest pterosaur ancestors were tiny, agile reptiles, and the earliest dinosaur predecessors likely were, as well. The "terrible lizards" and their relatives did not emerge from earlier Triassic giants, but arose from miniature creatures. Such small animals can sometimes have difficulty regulating their own body temperature, as they have a great amount of surface area compared to their internal volume. Not only do they have to eat a great deal, but shifts in outside temperature—a chilly day, or even too much sun—can have a greater effect on their internal heat. One way to cope with such shifts and maintain stability is through insulation. A coat of fuzzy, almost furlike feathers would both help keep body heat in during cold snaps and slow overheating in warmer spells. Direct fossil evidence will be required to test this hypothesis, and as yet no one has described a feathery fossil of a Triassic dinosaur, but the spread of the evidence suggests that dinosaur ancestors evolved fuzzy coats to help regulate their body temperature and this anatomical gift was passed on to the likes of the possible oldest dinosaur, the long-necked *Nyasasaurus*, and many dinosaurs that would follow.

Microraptor was not a bird, but a non-avian dinosaur that evolved a unique method of flying by using wings on the hind limbs as well as the arms.

In some dinosaurs, such as *Archaeopteryx* and *Microraptor*, ancestral feathers were modified for flight. Paleontologists often pay special attention to whether dinosaur wing feathers are asymmetrical or not, a shorter leading edge of the feather at the forefront of the wing being an aerodynamic adaptation that indicates powered flight rather than parachuting or gliding. More broadly, however, dinosaur feathers embodied many different uses in the day-to-day lives of the animals, which no doubt helps explain how widespread plumage became. Feathers kept dinosaurs warm, camouflaged them as they moved through prehistoric forests, and even impressed other dinosaurs. The banded tail of *Sinosauropteryx*, for example, was likely a social signal to communicate with each other the same way that cats and other mammals posture with their tails. The bristles of *Psittacosaurus* are more enigmatic, but could have rustled when shaken, or perhaps helped keep some of the sun's heat off the dinosaur's backside. The fuzzy coat of the 30-foot-long (9 m) carnivore *Yutyrannus*, by contrast, likely kept the dinosaur warm in a high latitude habitat that could have seen snow in the Early Cretaceous, and the fanlike feathers on the arms of the ostrich-like "bird mimic" *Ornithomimus* might have been used in courtship displays just like their modern counterparts do. Feathers did not evolve specifically for flight. Instead, flight became possible because dinosaurs had been evolving many different feather types and functions for tens of millions of years before *Archaeopteryx*. And as paleontologists have studied the fossil plumage, they discovered the hidden evidence to a mystery that experts were nearly certain would never be cracked. Feathers allowed scientists to begin reconstructing dinosaur colors.

The potential for the fossil record to preserve the original hues of once-living creatures has been known since at least the nineteenth century. In a letter dated December 9, 1833, the British paleontologist Elizabeth Philpot sent a letter to her friend Mary Buckland depicting the skull of a marine reptile, an ichthyosaur, that fellow paleontologist Mary Anning had found. Philpot had drawn the skull in ink scraped out of a fossil squid relative called a belemnite, still capable of acting as a pigment despite being Jurassic in age. The ink sac fossil was preserved delicately enough to include tiny, pigment-carrying organelles called melanosomes, whose shape and distribution in a tissue can make various shades from black and gray to rust and red.

Paleontologists who came after Philpot were aware that some soft tissue fossils seemed to enclose areas of microscopic structures, but these were often written off as fossilized bacteria. Bacterial mats (multiple resistant layers of microorganisms) often form around decaying carcasses, and so the ovals and spheres were thought to be these decomposition products. But while looking at a different fossil ink sac in 2006, paleontologist Jakob Vinther questioned the prevailing wisdom. Perhaps the microscopic structures really were melanosomes. The question opened up a series of studies on fossil color, including on a banded fossil bird feather that contained melanosomes in the darker portions but lacked them in the lighter parts. It seemed that the fossil record really could preserve clues to color, comparisons between fossil melanosomes and the tissues of living animals allowing researchers to effectively reverse engineer some prehistoric colors. And if the method worked on an isolated bird feather, then perhaps dinosaur feather colors could be reconstructed, too.

The earliest birds date to the Jurassic, meaning that avian dinosaurs thrived alongside other dinosaurs through the Cretaceous. The ancient bird *Confuciusornis* lived alongside dinosaurs from many other groups about 125 million years ago.

A different research team was the first to publish a paper on dinosaur color, focusing on the iconic *Sinosauropteryx*. The dinosaur appeared to have a coat of rusty feathers. Subsequent research has suggested that the dinosaur's color and pattern are an example of countershading, being dark above and light below, to help conceal the small carnivore in the ancient forests where it lived. Vinther and colleagues quickly published their own study on a different dinosaur, an *Archaeopteryx*-like dinosaur called *Anchiornis*. The researchers wanted to reveal the dinosaur's colors in their entirety, finding that *Anchiornis* may have looked something like a magpie—charcoal and white colors on the body with a burst of reddish feathers on top of the head. Since those first studies in 2010, experts have reconstructed the shades of several other dinosaurs, finding that *Microraptor* likely had a dark, glossy look like ravens and the "rainbow" dinosaur *Caihong* was colorfully iridescent, like a hummingbird or a pigeon's neck.

Not all dinosaur colors were made by melanosomes. The tiny structures allow paleontologists to reconstruct color because their shape and distribution reflect light back in different ways, which can be compared to melanosomes in living animals where the color is known. Other colors, such as bright oranges and yellows, are created by biochemical compounds rather than structures like melanosomes, and experts have not as yet found a reliable method to draw them out. Nevertheless, paleontologists never anticipated the ability to assess dinosaur coloration and what it might mean. The findings so far indicate that feathery dinosaurs wore colors similar to those seen among living birds, allowing for further research into what dinosaur colors meant for their behavior and natural history. The discovery of fluffy, fuzzy, and feathery dinosaurs not only confirmed that birds *are* dinosaurs, but have allowed paleontologists to study aspects of dinosaur lives that were long thought to be out of reach. Feathers didn't just change what we think dinosaurs looked like, but how the reptiles lived.

CHAPTER 7

Dinosaur Diets

From the moment of the large theropod's scientific debut, there was no doubt what *Megalosaurus* ate. A fossil jaw from the "great lizard," described by William Buckland in his 1824 write-up of the dinosaur, included a curved, serrated tooth jutting upward from the bony fragment, as well as several other tooth tips growing along the gumline when the animal had perished. The tooth was clearly meant to cut and slice, a piece of anatomy intended for carnivorous activities.

A theologian as much as a geologist, Buckland later mused on the perfection of the fossilized points in a series of books commissioned to highlight the goodness of the Christian god in creation. Buckland was responsible for the sixth of these Bridgewater Treatises, published in 1837 and focused on geology, in which he made special mention of *Megalosaurus* as a predator perfectly honed to kill without causing suffering. The teeth of *Megalosaurus*, Buckland wrote, "appear to have been admirably adapted to the destructive office for which they were designed." He likened the dinosaur's teeth to saws and pruning knives, the teeth presaging the tools humans would later invent to puncture and cut. So perfect were such teeth, Buckland mused, that they were clearly suited "materially to diminish the aggregate amount of animal suffering" through their efficiency.

Perhaps the Jurassic prey of *Megalosaurus* would differ from Buckland's interpretation. The dinosaur was certainly a carnivore, as Buckland and his peers immediately recognized, but there is no reason to believe that the dinosaur was especially artful or empathetic with its bites. The teeth and jaws of all carnivorous dinosaurs acted like enormous shears, each tooth adapted to puncture through skin and muscle. With teeth sunk into flesh, carnivorous dinosaurs used the powerful muscles of their necks to pull their heads back, the tiny serrations of each tooth easing each cut through the tissue. Whatever morsel the predator pulled off was then swallowed whole, leaving the digestive system to break down the muscles, viscera, and bone.

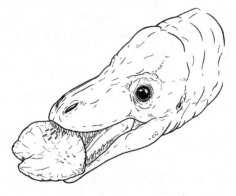

T. rex and other large, carnivorous dinosaurs frequently bit and consumed bone, which often passed into their feces. Fossilized dinosaur poop is called "coprolite."

Fossilized feces confirm what the anatomy of carnivorous jaws imply. Known as coprolites to experts, the preserved dinosaur dung often shield fossilized remnants of dinosaur meals and offer clues about how dinosaurs fed. In the case of big carnivores, a Cretaceous era plop from the most famous of all dinosaurs underscores the fact that predatory dinosaurs bolted down large chunks of food and probably had to eat often in order to fuel their warm-running bodies. The fossil, described in 1998 by paleontologist Karen Chin and colleagues, was 17 inches (44 cm) long and found in the Late Cretaceous rocks of Saskatchewan. More than 30 percent of the coprolite was bone fragments, and some tatters of striated muscle tissue could still be detected inside. Only one carnivore found in the same rocks could have left such a clue behind: *Tyrannosaurus rex*. The reptile had clearly torn off parts of a carcass and swallowed the pieces, its digestive system moving so rapidly that some pieces of the prey hadn't been fully digested. While it's possible that the petrified dung might have been unusual—perhaps the tyrannosaur had diarrhea—it's more likely that big, carnivorous dinosaurs were messy eaters who had no qualms about ingesting bone as well as soft tissue. The skeleton of another tyrannosaur—a young *Gorgosaurus* described by paleontologists in 2023—had the legs of two smaller dinosaurs inside its stomach. Whether the young tyrant hunted or scavenged its meal isn't clear, but the growing dinosaur prioritized tearing off the heavily muscled, nutritious legs of two smaller, parrot-like dinosaurs and swallowed the limbs whole, letting the stomach and intestines draw all they could from the meal. Carnivorous dinosaurs didn't use a varied dental toolkit of incisors, canines, and cheek teeth to snip off and break down carcasses like jaguars or wolves do. The anatomy of dinosaurs like *Megalosaurus* and *Gorgosaurus* were suited to ingesting huge quantities of meat as quickly as possible.

Rare fossils such as the *T. rex* coprolite and *Gorgosaurus* stomach contents have been essential for what paleontologists have previously assumed on the basis of teeth and jaws. In broad strokes, paleontologists have been able to look at dinosaur dentitions and get a rough idea of what the reptiles were eating. Carnivorous dinosaurs typically have sharp teeth suited to puncturing flesh, and are often—though not always—curved and serrated. Herbivorous dinosaurs from beaked *Iguanodon* to giant *Apatosaurus* had a wider array of teeth, and sometimes no teeth at all. Many plant-eating dinosaurs had teeth best suited to nipping off vegetation that was immediately swallowed, but some groups, such as the duck-billed and horned dinosaurs, evolved unique ways of chewing to begin to break down their meals before sending it along their digestive tracts. For many dinosaurs, it's relatively easy to take a quick look at their skeletons and quickly gauge whether they mostly ate plants, meat, or significant amounts of both.

Gut contents provide a direct record of what dinosaurs were eating just before death. Fossilized gut contents of the small, carnivorous dinosaur *Sinocalliopteryx* indicate that it fed on smaller feathered dinosaurs and birds.

Teeth can only take us so far, however, in approaching that classic dinosaur question so many of us have entertained in our imaginations and childhood sandboxes—*what did dinosaurs eat*? Gazing at the sharp teeth of *Allosaurus*, for example, we might imagine the carnivore in epic clashes with herbivorous giants such as *Brachiosaurus* or the spike-tailed *Stegosaurus*, when the predator more likely tried to single out juvenile dinosaurs and scavenged on the carcasses of multi-ton giants, dead from drought, disease, or some other cause. The 152-million-year-old Mygatt-Moore dinosaur quarry in western Colorado, for example, contains a large number of disarticulated and tooth-marked bones that indicate scavenging in a stressed environment. Even *Allosaurus* bones have carnivorous dinosaur toothmarks on them, either made by other predators or by *Allosaurus* with no qualms about cannibalism. Looking at a dinosaur's teeth can give us some idea of how they selected and manipulated their food, but the truth of what dinosaurs actually ate comes from toothmarks, coprolites, and other forms of trace fossils that reflect specific moments in dinosaur lives.

For the most part, paleontologists have found that meat-eating dinosaurs, both large and small, used their jaws in similar ways. A Jurassic bone scored by teeth, preserved in the quarry wall of Utah's Dinosaur National Monument, is a case in point. The bone itself belonged to a young sauropod dinosaur like *Diplodocus* or *Apatosaurus*, and it's the only bitten bone that paleontologists have ever found from the site. Three possible theropod predators could have left toothmarks like the long scratches on the bone—the huge, sharp-toothed *Allosaurus*, the horned *Ceratosaurus*, and the large, knife-toothed *Torvosaurus*. Despite the anatomical differences between these carnivores, however, paleontologists haven't been able to identify which predator is responsible for the bite marks. All three carnivores belonged to different groups of theropod dinosaurs, and all three had distinctive teeth that can be distinguished from the others, but they all used the same puncture-pull strategy to feed on carcasses, as all left the same damage patterns on bones. While this behavior frustrated paleontologists' efforts to work out what species bit the thigh bone before it was buried, the confusion indicates that carnivorous dinosaurs of different ancestries and sizes had settled upon a consistent, effective way of removing flesh. Some were even precise about defleshing carcasses. Large tyrannosaurs often had smaller, closely packed teeth at the front of their jaws, and one duck-billed dinosaur fossil found in the Gobi Desert showed signs that a tyrannosaur selectively scraped meat from a few parts of the skeleton. Even

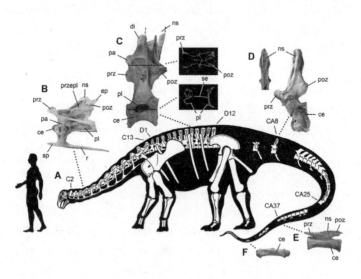

The sauropod dinosaur *Nigersaurus* had a skull that was especially adapted for eating large amounts of rough, low-growing vegetation such as horsetails and cycads.

though tyrannosaurs were capable of tearing off body parts and smashing bone into splinters, this one was scraping off pieces of flesh in a more delicate way. When you mostly interact with the world with your mouth, precision is an excellent skill.

Herbivorous dinosaurs were much more varied in how they fed. Sauropod dinosaurs, for example, had stubby teeth suited to clipping vegetation but were incapable of processing or mashing up that food. Plants are often fibrous, and debris like dirt that sauropods inadvertently ate with their Mesozoic salad undoubtedly scratched their teeth, too, and so the giant dinosaurs were constantly replacing their teeth throughout their lives. The spoon-toothed sauropod *Camarasaurus*, for example, replaced a tooth about every sixty-two days, while *Diplodocus* replaced a tooth about every thirty-five days. Some sauropods took this ability to even greater extremes. The boxy-headed cow-like *Nigersaurus* replaced a tooth about every two weeks, likely a consequence of feeding on low-growing, abrasive plants like horsetails growing along ponds and streams. And despite the fossil lore around these reptiles, most sauropods didn't intentionally swallow stones to grind up plants in their guts. Modern crocodilians swallow stones (which become known as gastroliths) to help maintain neutral buoyancy in the water, and many birds swallow stones to create a "gastric mill" suitable for grinding up nuts, seeds, and other hard foods, but paleontologists have never found solid evidence that sauropod dinosaurs regularly did the same. Polished stones found in the gut regions of sauropod skeletons are often rocks that have been transported by the waters that buried the huge animals, which tracks with how strong a stream or water flow would have to be in order to carry enough sediment to bury multi-ton dinosaurs. To date, the best evidence for gastroliths among dinosaurs that paleontologists have found are among small, toothless theropod dinosaurs that convergently evolved to behave like ground-dwelling birds. The toothless dinosaurs plucked up hard foods, like seeds and insects, which were soon busted up and broken down by the pebbles the dinosaurs had swallowed.

Most other herbivorous dinosaurs—from the spiky *Stegosaurus* to the beaky *Iguanodon* and more—ate with scissor jaws. What they fed upon depended on what they could reach. Sauropods were suited to reaching high and low, left and right, over a wide span. Other dinosaurs, such as the low-slung armored ankylosaurs, fed lower to the ground, or could reach only a little way above the soil. Relatives of today's ginkgoes, conifers, horsetails, ferns, and cycads made up a great deal of this vegetation, as the earliest flowering plants evolved about 125 million years ago but were relatively rare until after the asteroid-triggered disaster that ended the Cretaceous period. But some herbivorous dinosaurs evolved ways to get more nutrition out of their meals, and there are no better examples than the poorly named duck-bills.

In a superficial sense, dinosaurs like *Edmontosaurus* look a little duck-like. The Cretaceous dinosaurs had long, low, broad jaws that vaguely resemble the beak of a familiar mallard. But "shovel-beaked" is probably a better description of hadrosaur anatomy. A few specimens, such as the aforementioned baby "Joe" the *Parasaurolophus* and an *Edmontosaurus* skull on display at the Natural History Museum of Los Angeles, have natural molds of the rough beak that hung over the front of the skull. Hadrosaur beaks were not thin, close-fitting tissue structures, but thick, tough layers of keratin that angled downward in front of the jaw like shovels. More importantly, ducks don't have teeth but hadrosaurs certainly did. In the *Edmontosaurus*, for example, the dinosaur could have more than fifty columns of closely packed, lance-shaped teeth arranged in what paleontologists often call a tooth battery. It was a broad, rough grinding surface, and the rest of the dinosaur's skull was adapted to smear plant material across those constantly replaced teeth.

Hadrosaurs chewed in a different way than mammals do. When we chew, for example, our jaws open and close and even shift side-to-side some. We have specialized teeth, our premolars and molars, peaked with cusps that help crush and pulverize whatever we're eating, increasing the surface area of the food so it breaks down easier when it hits our stomachs. *Edmontosaurus* did something different. The joints between their skull bones had more flex to them than ours do, and so, when the dinosaur raised its lower jaw to meet the upper, bones in the skull shifted in response to the push from the lower jaw and this swung the tooth batteries of the upper jaw slightly to either side. When the lower jaw dropped again, the tooth batteries would go back to their resting position, ready to have the bottom jaw shove them again for the next chew. With a little anatomical adjustment, hadrosaurs were able to break down plant food in a way few other dinosaurs could.

We know hadrosaurs made good use of their impressive teeth because of coprolites, which have imparted some critical lessons about dinosaur ecology. Often, we think of dinosaurs eating a specific type of food. Herbivores would have snacked on ginkgo leaves and monkey puzzle tree branches, and carnivores loved to munch whatever flesh they could peel from a carcass. But we know from modern animals that these

DINOSAUR DIETS 121

rules aren't strict. Some herbivorous animals not only swallow insects with their plant meals, but sometimes eat small animals, too. Carnivores, also, sometimes eat fruits and other plants, either as part of the gut contents of their prey or just because the mood strikes them. Animals evolve to feed in particular ways, suited to everything from trees to other animals, but dinosaurs were not selecting specific, isolated ingredients to consume. In the case of the Late Cretaceous hadrosaurs, munching on a rotting log meant getting a little extra protein with the fiber.

While the eggs, embryos, and babies of the hadrosaur *Maiasaura* made the dinosaur famous, the reptiles also left behind pats of fossil dung in the eighty-million- to seventy-four-million-year-old rocks of Montana. Each one contains anywhere from 13 to 85 percent wood fragments. The dinosaurs were clearly not just nibbling leaves and accidentally chewing up some wood with the rest of their meals. The high volume of wood in the coprolites, paleontologist Karen Chin hypothesized in a 2007 study, is because the hadrosaurs were intentionally eating rotting conifer logs. The beaked dinosaurs were intentionally chomping into and chewing downed and decomposing parts of conifer trees, not just in a moment of stress but apparently year after year. Hidden morsels found within the dinosaur coprolites offer a clue as to why the hadrosaurs were intentionally seeking out punky wood.

Despite running through the bodies of *Maiasaura*, the digested wood in their coprolites still retained some woody details from when the fragments were first eaten. The wood wasn't fresh, but showed signs that fungi were growing on and inside the rotting logs. When the *Maiasaura* took a mouthful of decomposing log, they were also getting proteins from the fungus and other decomposers that were breaking down the wood from the inside. Coprolites found in similar-age rocks from Utah indicate that this was a common practice for hadrosaurs. Just like the coprolites from Montana, the Utah fossils were full of wood that had been rotting before consumption. Shells of small crustaceans also showed up in the coprolites, providing the dinosaurs with extra calcium and other nutritional resources. It may be, Chin proposed, that dinosaurs around nesting season were searching for extra nutrition to form their eggs, with rotting wood, fungus, and invertebrates being what hadrosaurs craved before building their nests. Especially during a time long before grasses and the spread of flowering plants, the way conifers and other ancient vegetation broke down likely provided rare, important resources for hungry hadrosaurs.

Of course, fossilized dung doesn't even have to leave the body to give paleontologists some insight into dinosaur diets. A small armored dinosaur called *Minmi* that lived in prehistoric Australia was found with digested gut contents that had already passed through the stomach but not exited the body, called a cololite. The mass was packed with vascular plant tissue, indicating the dog-size dinosaur was eating ferns, fruit that fell to the ground, and other low-growing vegetation. Even tinier fossils found in another armored dinosaur revealed what the herbivore was munching on prior to its death. Pollen is incredibly resilient in the fossil record, with a whole subscience—palynology—devoted to it. A cololite found inside the body cavity of an exquisitely preserved ankylosaur called *Borealopelta* not only included visible shreds of plant material, but fossil pollen that allowed more specific identifications to be made. (After all, it can be hard to reconstruct what someone was eating after it's already been run through half the digestive system.) Paleontologists found fifty different pollen types inside the dinosaur's gut contents, primarily from ferns, clubmosses, and liverworts, along with some conifers and a few flowering plants. The forests where *Borealopelta* lived were apparently carpeted by mosses, ferns, and other plants that love moist habitats, and the dinosaur snaffled up whatever its boxy snout could reach—including some burnt wood, given the charcoal in the gut contents.

Whether it's fossil pollen or bones inside of stomach cavities, prehistoric poop, or other fossils, evidence of dinosaur diets tells us far more than what a particular animal consumed. Often it's difficult to tell for sure which dinosaur species lived side by side and which were split by a million years, just as we can easily pick out the sharp teeth of a carnivorous dinosaur but be uncertain what sorts of meat that predator preferred. Each dietary connection that gets made, then, begins to build out the dinosaur's ecology. The clues give experts a sense of what individual dinosaurs were doing in their last days, how dinosaur diets changed with age or with seasons, and what sorts of plants grew in the area that otherwise might not be preserved. Dinosaur gut contents, especially, fold other pieces of their prehistoric habitats into the fossil record, preserving parts of ancient ecology that would otherwise be entirely obscure. Evidence of dinosaur diets were not just a matter of "you are what you eat," but hints about how the world worked long before the interglacial ecosystems we're familiar with today changed our environment forever.

CHAPTER 8

How Dinosaurs Made
Their World

If decades of dinosaur media is to be believed, Mesozoic reptiles liked to live in sweltering jungles. From 1940's *Fantasia* to 1993's *Jurassic Park* and its growing number of sequels, dinosaurs are envisioned as being most comfortable in habitats flanked by palm trees and hemmed in with thick-leaved tropical plants. The visions seem to perfectly suit the reptiles, giving them plenty of places to lurk and sometimes crash through the thick undergrowth. But dinosaurs never lived in such rainforests. Such dense growth wasn't even possible during the heyday of the dinosaurs. The fact that thick tropical rainforests began growing after the dinosaurs is an indication of just how heavily dinosaurs influenced Earth throughout the Mesozoic era.

Earth's first tropical rainforests didn't appear until about sixty million years ago, after the asteroid strike that ended the Cretaceous period. Some of the reasons have to do with the effects of the disaster, such as iron from pulverized rocks being spread across the planet in the impact plume and the fact that angiosperms, or "flowering plants" like beans, coffee, peppers, palms, and more, were relatively rare before the event but were able to grow quickly afterward. But dinosaurs themselves were another reason. Even though there were plenty of small and medium-size dinosaurs all over the planet at the end of the Cretaceous, giant, multi-ton dinosaurs like *Edmontosaurus* and *Triceratops* were still numerous and widespread at that time. These dinosaurian megaherbivores not only ate plants, but pushed over trees, trampled the ground where they walked, and left seed-filled dung in their wake, creating relatively open forests with plenty of dinosaur-size gaps in them.

The herbivorous *Maiasaura* had powerful jaws capable of reducing rotting logs to pulp, a behavior that played an important role in the dinosaur's ecosystem.

When paleontologist Karen Chin described fossil *Maiasaura* dung from Montana, part of the reason she was able to do so was because the coprolites had been burrowed into by small invertebrates. The little tunnels had been made in the Cretaceous and infilled with sediment when the fossils were buried. The dinosaur droppings didn't just reveal what the hadrosaurs were eating, but connected them to their broader habitat. The landscape *Maiasaura* inhabited was dotted with conifer trees, likely with game trails made through the stands of trees where the 4-ton (3,600 kg) herbivores walked. When those trees fell, they attracted a host of different decomposers from fungi to snails. *Maiasaura* fed on these rotting longs, even though the bacteria in their digestive systems could not fully break down the woody material. The remains were deposited back onto the soil where they became a new resource for all sorts of invertebrates, including beetles that both ate the dung and used it to incubate their young just like many dung beetles today. Dinosaurs were not isolated from the other creatures of their time, but participated in vast, intimate food webs that created landscapes of familiar components but still seem strange to us from our modern standpoint.

Just like animals alive today, dinosaurs both evolved according to the ever-changing conditions of our planet while also influencing life on Earth themselves. The Mesozoic Earth was not merely a diorama that dinosaurs existed within as climates and vegetation changed around them. Dinosaurs can be understood only in the context of their paleoecology—how they connected to the natural world, from the parasites that lived on their bodies to influencing plant species that would evolve defenses against being eaten. Dinosaurs even changed the planet in a more literal way. A significant part of the dinosaur fossil record is how the reptiles altered the topography of the places they lived.

Everywhere they went, dinosaurs left footprints behind. Some of these became preserved as fossils, with clean, distinct footprints often gaining the most attention. But sometimes dinosaurs trampled a place so significantly that paleontologists can detect the ways the many scaly feet churned the soil. Experts call this phenomenon dinoturbation. A Cretaceous footprint site in Brazil's Araripe Basin is a standout example

Footprints of sauropod dinosaurs found in Australia indicate how the reptiles trampled low-lying areas and changed them into lagoons. Where dinosaurs chose to walk could significantly alter a landscape over time.

of what a repeatedly dinosaur-stomped surface looks like. Documented by paleontologist Ismar de Souza Carvalho and colleagues, large dinosaurs walked in some ancient areas so frequently that the ground beneath their feet began to deform and create depressions. Over time, these shallow bowls began to deepen enough to collect water, becoming puddles, then lakes which accumulated compounds like sodium carbonate to such a degree that they became strange soda lakes. A similar phenomenon has been documented in western Australia, as well, where the hefty steps of sauropod dinosaurs altered the thin layers of sediment beneath each imprint to create a landscape scattered with lagoons. Dinosaurs didn't rule the Earth. They inadvertently shaped it to their needs.

It wasn't only the giant dinosaurs that affected ecosystems, of course. The fine-scale preservation that has allowed many feathered dinosaurs to enter the fossil record with their plumage intact also ensured some of their meals were fossilized. Skeletons of the early bird *Jeholornis* from the Early Cretaceous rocks of China, for example, have been found with clusters of seeds inside their body cavities. The birds were not plucking the seeds and eating them one by one, but rather were eating fleshy fruits with dense clusters of seeds inside. Additional evidence of phytoliths—or microscopic, silica-rich structures found in plant tissues—indicates *Jeholornis* was eating the leaves and fruits from early magnolia relatives around 125 million years ago, demonstrating that early birds were fluttering through Cretaceous forests above the heads of other dinosaurs in their search for fruit and leaves. This specialization likely had significant consequences for the prehistoric forests the birds resided in. Rather than destroying or grinding the seeds, *Jeholornis* swallowed them whole or encased in fruit. The seeds took a ride through the bird's digestive system only to be deposited in a new spot, along with some fertilizer. Repeated over and over again, the ingestion and deposit of these seeds consequently helped the seeds of the trees, undoubtedly changing the nature of the forests *Jeholornis* lived in. Even if only relatively few trees sprouted and grew from this dinosaurian assistance, the birds were still planting more of the trees they liked to eat and affecting the forest's very nature.

Why *Jeholornis* would specialize in treetop plants is another clue to ancient habitats, a phenomenon ecologists call niche partitioning. When multiple, similar species live in the same space, they are likely to vie for the same food and resources, seeking to survive shoulder-to-shoulder. Direct competition can be stressful and even lead some species to be edged out by others. But even among extremely similar creatures, small variations open different possibilities and can lead different species to open their own niches where there is less direct competition for what they need. The various lemurs of Madagascar are a potent modern example—multiple primate species living in the same forests but preferring different food and living at different heights—but paleontologists have seen this pattern in dinosaurs, too. *Jeholornis* was a plant-eating bird that lived alongside others that consumed insects or small vertebrates like lizards, partly explaining why the deposits of the Jehol biota, made up of Cretaceous fossils of northeastern China for which the bird is named, boasts so many feathery dinosaur species. And at a larger size, the dinosaurs found in Alberta's Dinosaur Provincial Park are another demonstration of how dinosaurs made their own space on the landscape.

The ankylosaurs, ceratopids, and hadrosaurs of the Dinosaur Park Formation fed on different plants and at different heights, allowing many giant plant-eaters to coexist.

During the Late Cretaceous, about seventy-five million years ago, at least three different forms of giant, plant-eating dinosaurs lived among the floodplains and forests of what's now Alberta. The heavily armored ankylosaurs shuffled along next to horned ceratopsids and the shovel-beaked hadrosaurs, all of which would have been after similar plant foods. Based upon skeletal anatomy and how far dinosaurs in each group could stretch their necks, both the ankylosaurs and ceratopsians had to focus on whatever food they could get within three feet of the ground. Ferns, cycads, young trees, and other low-growing plants were their staples, with the ankylosaurs preferring softer foods and the ceratopsids better equipped to munch on tougher plant material. Hadrosaurs, however, could feed up to 6 feet (1.8 m) off the ground when on all fours and more than 16 feet (5 m) off the ground when standing on their hind legs, a significantly greater reach that would have enabled them to chew on tree branches and the taller parts of shrubs. The specifics of what different species ate awaits more direct evidence, but the fact that ankylosaurs, ceratopids, and hadrosaurs are often found together in Cretaceous rocks across North America suggests that the feeding abilities of these reptiles allowed them to coexist by preferring different offerings from the same habitat. Multiple dinosaurs fed at different levels of the forest, affecting the evolution of various plant species as they did so.

The various ways that dinosaurs interacted with the environments they called home is a relatively new area of study, one that was reliant on new fossil finds in the latter part of the twentieth century, as well as technological advances such as CT scanning that's allowed experts to analyze tiny fossil details that were previously entirely beyond the reach of the imagination. But there is one aspect of dinosaur ecology that paleontologists have been pondering ever since they recognized that an especially long "Age of Reptiles" preceded the "Age of Mammals." The earliest mammals originated during the Mesozoic, right alongside the evolutionary explosion of reptiles. These ancient beasts seemed to be insectivores and universally small, with big eyes that suggested they were active at night. Paleontologists pointed out that not until after the

Cretaceous did mammals get larger than the size of a house cat. Dinosaurs seemed to keep mammals in submission, our ancestors and relatives squeaking out a living under the feet of the "terrible lizards," until a great extinction cleared the way for our forebears.

While twentieth-century experts and science communicators depicted dinosaurs as sluggish, dim-witted, and even freakish, our early mammal relatives were thought of as evolutionary underdogs. Mammalian chauvinism led paleontologists to cast small, beetle-crunching mammals of the Mesozoic era as superior creatures that were being held down by the tyranny of cold-blooded reptiles, an antagonistic relationship that even spun off into a proposed reason for dinosaurian extinction. The Triassic, Jurassic, and Cretaceous worlds were rife with dinosaur eggs that were thought to be left unguarded by careless parent reptiles, perfect meals for mammals that were likely too small to be noticed. Perhaps early mammals staged a kind of prehistoric coup, one oft-repeated hypothesis suggested, growing in number as they ate nest after nest of dinosaur eggs until the great reptiles could no longer reproduce fast enough to sustain their numbers. Dinosaurs were treated as both fascinating but repulsive, bizarre creatures that got in the way of mammalian ascent and paid the ultimate price for it.

Even after the death-by-omelet hypothesis faded out of favor, as it could potentially explain only why dinosaurs disappeared and not any of the other forms of life that vanished at the end of the Cretaceous, the image of Mesozoic mammals as meek, nocturnal, and downtrodden hung on. The idea undoubtedly spoke to our own modern-day feelings about dinosaurs, creatures that were fascinating but that we also spoke of as "dominating" and "ruling" the Earth. The narrative for our ancestors was one of subjugation by the reptiles, with an emphasis on large body size as superior seeming to add credence to the notion that mammals couldn't really thrive until the big dinosaurs were gone. A few rare fossils even seemed to speak to this idea, like dinosaur claw marks preserved around fossil mammal burrows found in Utah—evidence of a raptor-like dinosaur scrabbling into the ground to try and nab whatever mammals might be hiding inside.

But just as paleontologists had underestimated dinosaurs for decades, so, too, had they misunderstood Mesozoic mammals. The supposed antagonism between dinosaurs and mammals has been overplayed, and in some ways misunderstood because of a focus on competition for ecological prominence. Mammals of the Triassic, Jurassic, and Cretaceous were not all shrewlike creatures that ate bugs and feared reptilian talons. Mammals can truly be said to have thrived during the Mesozoic, and new finds are continuing to alter the picture.

A fossil of the badger-like mammal *Repenomamus* was found preserved with the small, horned dinosaur *Psittacosaurus*. The mammal was biting the dinosaur when the two perished and were buried.

Paleontologists often debate whether "mammal" should refer to what are often termed crown mammals— defined by the last common ancestor of monotremes, marsupials, and placental mammals—or include all the forms more closely related to mammals than reptiles, which are broadly called mammaliaforms (of which crown mammals are a subset). In a general sense, though, fuzzy creatures that we would likely call a mammal if we saw one alive were scurrying around by the Triassic, 225 million years ago, at the latest. The beasts flourished among the forests, floodplains, and deserts of the Mesozoic world, spinning off a variety of different forms. There were ancient mammal equivalents of squirrels, shrews, otters, aardvarks, flying squirrels, and more. And some of them were demonstrably not afraid of dinosaurs. A fossil of the badger-size mammal *Repenomamus* described in 2005 was found with baby dinosaur bones in its stomach, and another *Repenomamus* described in 2023 was biting a small horned dinosaur called *Psittacosaurus* in the ribs when both were suddenly overcome by a lahar debris flow of volcanic material and preserved. A focus on body size had misled paleontologists and caused dinosaur experts to overlook how varied Mesozoic mammals had become.

The emerging picture is that competition between different forms of early mammals restricted the evolution of our ancestors, not the dinosaurs. During most of the Mesozoic, it was relatively archaic forms of mammals—the mammaliaforms—that evolved into an array of different shapes and niches. The early beasts effectively took up a great deal of ecological space and their lineages held onto those spaces even as crown mammals emerged. It was only during the Cretaceous, as the older mammaliaforms began to disappear, that the

Interactions between mammals and their more archaic relatives, mammaliaforms (such as the shrewlike *Agilodocodon*), were more important to mammal evolution than competition with dinosaurs.

ancestors of today's marsupial and placental mammals began to evolve in new ways and establish the basis for their great flourishing in the Cenozoic era, or from the asteroid impact sixty-six million years ago to the present day. The story played out in parallel with the comings and goings of different dinosaur groups, having more to do with interactions between mammal lineages than supposed conflict with the dinosaurs. And even though all Mesozoic mammals found so far haven't been any bigger than an adult house cat, the fossil record may still hold some surprises. In 2019, paleontologists named an enormous protomammal from 205-million-year-old rocks in Poland, therefore living alongside the dinosaurs. It belonged to a tusked, piglike group called dicynodonts, but it grew to be more than 5 tons (4,500 kg) in weight and lived alongside early dinosaurs and other reptiles. The fact that such a large protomammal not only survived so long into the "Age of Reptiles," but had gone unrecognized until well into the twenty-first century, hints there may be other pockets in Mesozoic time when mammals and their relatives grew larger than paleontologists have expected. After all, the vast majority of mammal species alive today are small—the size of rodents such as mice and squirrels. The same was true in the Mesozoic. The diverse forms mammals evolved into, and the various niches they opened, are signs of success and not suppression by reptiles.

The legend, therefore, isn't true. Dinosaurs did not rule anything. The reptiles were part of an ever-changing world, influenced by both other living things and abiotic factors such as climate and rainfall patterns, while dinosaurs themselves changed the ecosystems around them. What dinosaurs ate and where they walked changed the world, just as the reptiles were shaped by the vegetation around them and the prehistoric climates. Their paleoecology is so often what we imagine when we try to picture how dinosaurs fit into and moved through their ancient environment, a shifting mural of prehistoric life that experts are only just beginning to fully perceive.

CHAPTER 9

Decoration and Defense

Fossil horns found in Colorado were initially mistaken as those of a bison before being recognized as those of the horned dinosaur *Triceratops*.

When nineteenth-century paleontologist Othniel Charles Marsh first saw the horns of the *Triceratops*, he didn't really know what he was looking at. The fossils had been found near Denver, Colorado, in 1887, accessible enough by rail that the find was shipped to him at Yale University in Connecticut. The horns were heavy and much of the original tissue had been replaced with minerals, clearly indicating that the animal had lived during an ancient time, but Marsh thought that the rocks the horns were found in dated back to the Pliocene epoch, the time period before the latest round of

Earth's Ice Ages, around 2.58 million years ago. With so little of the fossil record known, Marsh could go with only what he knew. The horns looked bison-like, and so he named them *Bison alticornis* that same year.

Marsh's mistake partly stemmed from the fact that when the horns reached him in New Haven, paleontologists had not yet recognized horned dinosaurs as a particular group of "terrible lizard." The numerous horned dinosaur fossils and bone beds that would soon be found in Montana, Wyoming, and Alberta had not yet been found, the Cretaceous fossils silently waiting for paleontologists to catch up. Only bits and pieces had been found so far, given different names but generally appearing to be enigmas. Even after Marsh himself recognized horned dinosaurs and named them ceratopids in 1888, he still maintained that the Denver-area horns were from a bison. He didn't revise his opinion until paleontologist John Bell Hatcher turned up a complete skull of a three-horned dinosaur in Wyoming, an animal Marsh named *Triceratops horridus* after the anecdote that the cowboy who discovered it was frightfully shocked by the skull's appearance.

Triceratops, of course, was a three-horned herbivore, the back of the skull supporting a solid frill of bone dotted with additional, triangular horns. When early paleontologists looked at the fossil reptile and its growing array of relatives, like the many-horned *Styracosaurus* from Alberta, they saw weapons. The horns of these dinosaurs must have worked as lances or spears, their frills acting as shields that protected the vulnerable neck from the bites of ravenous tyrannosaurs. The fact that such dinosaurs were found alongside some of the largest carnivorous species then known repeatedly led paleontologists to imagine fantastic clashes between the likes of *Tyrannosaurus* and *Triceratops*, the herbivore with their head down and threatening the voracious predator with their horned impalement aimed directly at the gut. Such monstrous and supposedly dim-witted reptiles must have fought each other regularly; after all, herbivores were thought of as slow-moving creatures that required defenses to keep carnivores from waddling up and chewing on their backsides.

Despite the fervent belief that *Triceratops* used their impressive horns to impale or otherwise ward off predators, paleontologists never found evidence of such a titanic clash. No *Tyrannosaurus* fossil has shown injuries definitively caused by a *Triceratops*, and all the tyrannosaur bite marks on adult *Triceratops* skeletons attested only to how the giant carnivore fed, mostly likely on individuals that had died from other causes. No matter how intuitive it seemed, the idea that horned dinosaurs evolved armaments principally to ward off predators did not hold up. Instead, the horns were more greatly influenced by social interactions—from showing off to combat.

Triceratops is relatively unusual among horned dinosaurs for having prominent brow and nose horns, but much more subdued ornaments on the rest of the skull. The dinosaur seemed like it would be capable of locking horns in combat, much like elk do with their antlers or elephants with their tusks. Cretaceous combat certainly would have left injuries on *Triceratops* skulls, and, when paleontologists looked at the *Triceratops* skull bones that would have been most likely damaged by such fights, they found exactly that. At least ten bones from the side of *Triceratops* frills, the squamosals, had damage, and at least seven cheekbones, or jugals, were injured, as well. And while some spiked *Centrosaurus* fossils had injuries in these spots, the count was significantly lower. *Triceratops* was clearly doing something different, the counted pathologies most consistent with horn-locking than battles with *T. rex*. The sharp points certainly could have given carnivores a second thought, but it was the social lives of *Triceratops* themselves that molded their "three-horned face" skull shape.

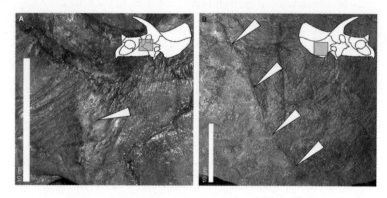

Injuries on multiple *Triceratops* skulls indicate that these herbivores regularly locked horns in combat.

The array of horned dinosaurs paleontologists have discovered since Marsh named ceratopsids in 1888 underscores the point. The elaborate headgear of horned dinosaurs is so distinctive that experts distinguish one species from another on the basis of their horn arrangements. Ceratopsids such as *Centrosaurus*, *Chasmosaurus*, *Triceratops*, and newer finds like *Kosmoceratops* and *Medusaceratops*, had horns on their noses, over their eyes, on their cheeks, and around the borders of their frills, all in different shapes and orientations. If these dinosaurs all evolved their horns to principally ward off predators, we'd expect those interactions to hone ceratopsid horns down to a small range of expressions over the millions of years. There would likely be an ideal configuration for telling tyrannosaurs to back off. Instead horned dinosaurs present an array of different horn shapes, sizes, and orientations over time, hinting that interactions between individuals of the species were a central factor in how these dinosaurs evolved. Within the past twenty years, paleontologists have stopped taking defense as a default explanation and have instead started asking whether so many seemingly bizarre dinosaur features might have been shaped by other evolutionary pressures.

Part of the misapprehension that's surrounded bizarre dinosaur features has had to do with what paleontologist Stephen Jay Gould termed "adaptationism." When biologists look at a creature living or extinct, he noted, there's a tendency to think about the function of each part and how that function created the form. One of the classic examples Gould pointed out are the conspicuously small arms of *T. rex*, which had attracted a number of largely untestable hypotheses ranging from hooking into prey to holding onto mates during copulation. If experts could figure out what such tiny arms were used *for*, then their evolution might consequently be explained in the sense of "this body part evolved for this specific function." But living things must be understood as wholes, as part of their environments, so isolated anatomy alone is an unreliable guide. Body parts often have multiple functions, just like our bones give us internal structure but also house marrow that creates red blood cells, or how the big ears of African elephants are both social signals and assist with helping the mammalian giants cool off thanks to the broad surface areas being good for releasing pent-up heat.

DECORATION AND DEFENSE

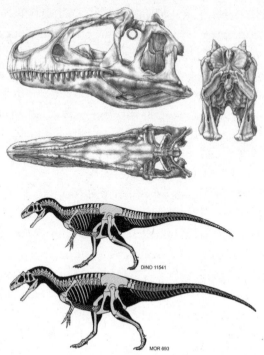

Many carnivorous dinosaurs, such as various *Allosaurus* species, had distinctive horns. These were not used as weapons but as social signals for picking mates and recognizing other individuals.

Just as paleontologists looked at the arms of *T. rex* and thought in terms of a single, primary function, so too did they traditionally look at horns, spikes, plates, crests, and other dinosaur features in terms of their main use. Nature is not so straightforward, and as experts have dug back into dinosaur biology, they've found how many elaborate bits of dinosaur anatomy served multiple functions in the lives of the animals. Interactions between members of a dinosaur species, as well as more antagonistic encounters between predator and prey, influenced what the reptiles became.

The evolutionary history of ankylosaurs—often called the armored dinosaurs—present a nuanced example of how strange dinosaur traits are shaped by multiple influences throughout the group's history. The herbivores were not merely "living tanks" meant to be impervious to carnivorous jaws. The osteoderms, or bones that grow in the skin, dotting their bodies could also help them recognize others of their species at a distance, impress mates, mobilize calcium for egg-laying, and help regulate temperature. The multifunctionality of ankylosaur armor explains its longevity.

Technically called thyreophorans, the earliest armored dinosaurs evolved in the early days of the Jurassic. They were small dinosaurs, about the size of a medium dog, like little *Scutellosaurus* from the two-hundred-million-year-old rocks of Arizona. The herbivore's back was dotted with rows of tiny osteoderms, and it seems more than a coincidence that some of the first large carnivorous dinosaurs are found in the same rocks. The movie-famous, double-crested predator *Dilophosaurus* lived among the same desert habitats, and was about five times longer than *Scutellosaurus*. While the skin bones didn't make *Scutellosaurus* impervious to big bites, they might have been enough to give the dinosaur a better chance at survival or perhaps even encourage *Dilophosaurus* to seek out less crunchy prey.

Osteoderms were far more than a physical barrier to teeth, of course. Studies of thyreophorans and living animals with osteoderms, like alligators, have indicated that the spikes, plates, and other ornaments of these dinosaurs carried out multiple functions. When female alligators are laying eggs, for example, their bodies resorb some calcium from their osteo-derms to embed in their eggshells. Studies of crocodilians and the plate-backed *Stegosaurus*, too, indicate that osteoderms can help reptiles disperse excess body heat. Blood vessels running along the bones can shunt warmer blood to the surfaces, allowing it to radiate off and cool the body. The thermoregulatory abilities of osteoderms, especially, became important as thyreophorans grew in size through the Jurassic. By the Late Jurassic, about 150 million years ago, the armored dinosaurs had flourished into the 10-foot-long (3 m), armor-covered ankylosaur *Mymoorapelta* and the 25-foot-long (7.5 m), plate-backed *Stegosaurus*. Some of them, like the extra-spiky *Stegosaurus* relative *Kentrosaurus* from the Late Jurassic rocks of Tanzania, could swing their tails so fast that their spikes could pierce hide, muscle, and into bone, if necessary, and the pebbly armor on the throat of stegosaurs suggests the beneficial defense against carnivorous teeth. The variety of overlapping osteoderm functions made thyreophorans a mainstay on Mesozoic landscapes from the Jurassic onward.

The spikes and ornaments of armored dinosaurs like *Ziapelta* were molded by social interactions and combat as well as for defense against predators.

The giant ankylosaurids of the Late Cretaceous, like *Ankylosaurus* itself, took some of these armor variations to the extreme. Spiky armor ran in rows down their backs and jutted out from their shoulders, their skulls decorated with various points and protrusions. One species found in the seventy-five-million-year-old rocks of Utah, *Akainacephalus*, was named for the supreme spikiness of its skull. And of course, ankylosaurids had specialized clubs of bone at the ends of their tails held aloft on an interlocking "handle" of V-shaped vertebrae.

It's not difficult to imagine a dinosaur like *Ankylosaurus* positioned with their belly low to the ground, swinging their tail to and fro to ward off a curious tyrannosaur. Paleoart is rife with such images, and there are a few fractured tyrannosaur shin bones that might have been the result of getting too close to a desired spiky meal. But while the coincident evolution of armored and carnivorous dinosaurs suggests that osteoderms really were important in defense and intimidating potential predators, there's no direct evidence that ankylosaurs were regularly bashing carnivore shins. Even though many carnivorous dinosaurs have multiple broken bones and pathologies in their skeletons—being a predator is dangerous work—there's no consistent signal that ankylosaurs were regularly breaking ankles. Instead, it seems that *Ankylosaurus* and kin were using their tail clubs on each other.

Just like the analysis of damage *Triceratops* inflicted on each other, the realization that ankylosaurs were bashing one another with hefty clubs of bone came from a recent find. Described in 2017, *Zuul* was a wide, spiky ankylosaur found in the seventy-five-million-year-old strata of Montana, one of the best-preserved armored dinosaurs ever found. When paleontologist Victoria Arbour and colleagues examined

The domed skulls of *Pachycephalosaurus* and related dinosaurs sometimes show injuries that may have come from headbutting in combat. These strikes were probably not head-to-head but directed toward the sides of the skull and the flanks.

the dinosaur's skeleton, they found the bony armor along the dinosaur's flanks and hips had been damaged. The distribution of the divots and breaks was consistent with what paleontologists would expect of two *Zuul* standing side-by-side, facing opposite directions, battering each other with their impressive tails. (If the damage had come from other causes, such as predators, falling trees, or tumbles, then the injuries would have had a broader and more random distribution over the body.) Looking more broadly to other ankylosaurs with tail clubs, Arbour and colleagues found that the dinosaurs did not start growing formidable, full-size tail clubs until sometime in their subadult years, hinting that the feature was more of a social feature than an essential defensive weapon. Perpetually cast as not-very-bright, fern-snuffling herbivores, the findings suggested that ankylosaurs had a broader range of behaviors than paleontologists assumed, engaging in social moments that certainly left their marks on the spike-studded dinosaurs.

The more paleontologists go back to the bizarre features of many dinosaur species, the more they're finding signs of social interactions driving elaborate anatomy rather than attack and defense. Part of the reason the pattern eluded experts

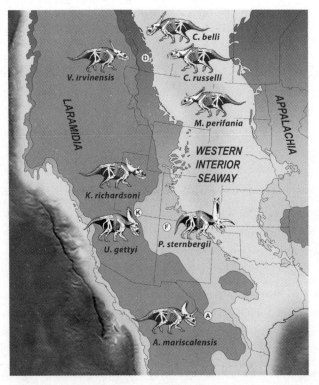

Many different horned dinosaurs lived up and down the North American subcontinent Laramidia during the Late Cretaceous. The stark differences in horn arrangements among neighboring species hints that they served more of a social function than a defensive one.

for so long is that, to date, no species of non-avian dinosaur has been conclusively shown to have distinctive, consistent anatomy tied to particular sexes. In living creatures, biologists can usually get a sense of whether different sexes of the same species vary in size, color, ornamentation, and other features, whether it's length of the canine teeth, body size, brighter colors during certain times of the year, or feathers meant to impress potential mates. The differences are marks of sexual selection, or evolutionary pressures related to attracting mates in one way or another. The peacock's tail is the standard example, birds with long trains of eye-spot feathers displaying to attract mates that lack the broad blue and green fans. So far as paleontologists have been able to tell from dinosaur skeletons, though, different dinosaur sexes were essentially the same at the bone level.

The vast majority of what we know from non-avian dinosaurs comes from skeletons, and by far most non-avian dinosaur species are known today from only parts of a skeleton or two rather than population-wide samplings. Even among non-avian dinosaurs that are recognized from dozens of skeletons, and from the same time and place, paleontologists have not found any positive evidence that one sex of a dinosaur species would be larger than another, or that the horns of one *Triceratops* sex would curve in a different way from its counterparts. Such sexual differences had long been treated as the hallmarks of sexual selection, and so paleontologists turned to other explanations for weird dinosaur headgear and lavish ranges of spikes. But in considering dinosaurs such as *Zuul*, "Joe" the baby *Parasaurolophus*, and others, paleontologists are beginning to track how bizarre dinosaur features changed through the animals' lives, and how it often took dinosaurs longer than expected to grow into their ornamentation. The ways dinosaurs interacted with each other drove a great deal of their anatomical weirdness that we so love when we visit museums, long lost moments of dinosaur behavior shaping so much of what the impressive reptiles came to represent.

CHAPTER 10

The Social Dinosaur

The American Museum of Natural History originally intended to use two *T. rex* skeletons to mount a standoff over a hadrosaur carcass.

The American Museum of Natural History had big plans for their two *Tyrannosaurus rex* skeletons. At the beginning of the twentieth century, fossil collector Barnum Brown had found two impressive specimens in Montana, the second even better than the first. To have multiple representations of a new dinosaur, the largest carnivorous reptile known, offered a rare opportunity. In 1913, museum paleontologists began making plans to mount the two partial skeletons, as well as the bones of a prey species, in a dramatic re-creation of a Cretaceous face off. Laid out in a miniature, each real and facsimile bone accounted for by artist Erwin Christman, one of the tyrannosaurs would crouch down with its jaws agape at a second, towering *T. rex* striding toward a duck-bill carcass between them. The diorama was an encapsulation of how dinosaurs were seen at the time, both powerful and sluggish, behaviorally complex but driven only by base instinct. After all, a dinosaur whose name meant "tyrant lizard king" would require a striking presentation for visitors.

The 1913 reconstruction never came to be. The museum opted to mount the second skeleton, the one with the better skull, instead, and the first, less-complete specimen was sold to the Carnegie Museum of Natural History in Pittsburgh in 1941 just in case World War II bombing raids hit the US East Coast. (The Carnegie would later create their own version of the 1913 model, having their *T. rex* face off against another in a mount unveiled in 2008.) Even so, the American Museum of Natural History's original plans hinted at a possibility that would lay dormant in paleontology until the 1970s. Despite the tail-dragging and cold-blooded plodding, the intended reconstruction presented *T. rex* as social animals that must have interacted with each other on the ancient landscape as well as their hapless meals. Whether relatively solitary or gregarious, dinosaurs must have had social lives.

In the case of *T. rex*, whether the carnivores were solitary for most of their lives or formed social groups is still a mystery. Earlier tyrannosaur species seemed to gather together at least sometimes, as shown by a trio of side-by-side tracks found in the roughly 70-million-year-old rock of British Columbia, but such trackway evidence has yet to be found for the largest tyrannosaur of all. But, at the very least, we do have evidence that *T. rex* meetings were not always friendly. Multiple *T. rex* individuals have healed bite marks on their skulls, including the best fossil of an adolescent *T. rex* yet known.

152 THE SHORTEST HISTORY OF THE DINOSAURS

"Jane" the *T. rex* was eleven years old when they died, just on the cusp of a teenage growth spurt that would see them become a multi-ton predator. Despite Jane's young age, however, the fossil had injuries that had previously been seen only on older *T. rex*. One of Jane's upper left jaw bones, the maxilla, has four puncture marks in it. Only another young *T. rex* could have left such damage, judging from the placement of the wounds. Jane had fought with another youngster some time before death, long enough to heal, just like older *T. rex* individuals. Similar damage has been found on the skulls of other carnivorous dinosaurs, too, such as the *Allosaurus* relative, the large theropod *Sinraptor*, and paleontologists have noted that living alligators and crocodiles bite each other on the face while fighting, too. For hundreds of millions of years, archosaurs have repeatedly turned to face-biting to solve their confrontations.

Not that every meeting of carnivorous dinosaurs led to bloodshed. After all, dinosaurs of every species would have to momentarily get along in order to create the next generation of eggs. Until recently, the fossil record has been near-silent on how dinosaurs would have impressed each other enough to mate (not to mention that the mechanics of dinosaur mating itself is something that remains in the realm of hypothesis more than hard evidence). So far, no dinosaur has ever been found preserved in the middle of a courtship display, and paleontologists have been largely left to guess how amorous dinosaurs behaved. But trace fossils left by dinosaurs have at least given us a peek at what must have been loud, smelly, and ridiculous scenes of Mesozoic romance.

In 2016, paleontologist Martin Lockley and a multi-institution collection of colleagues described what the researchers called "display arenas" made by large theropod dinosaurs akin to allosaurs or tyrannosaurs. Paleontologists found shallow, bowl-like depressions up to 6 feet (1.8 m) wide marked by long scratch marks, not unlike those made by some ground-dwelling birds today. The dinosaurs may have been acting like modern-day ostriches, which display and make long scrapes where they intend to nest. Whether the large dinosaurs that made the Cretaceous scrapes fluffed their feathers, made noises, shook their tails, or otherwise tried to show off, we may never know, but the broad dispersal of their scratches through Cretaceous rocks in Colorado, Canada, South Korea, and likely additional, yet-to-be reported sites indicate that theropods big and small have been strutting their stuff by scratching for millions and millions of years.

Some theropod dinosaurs scratched at the ground in vast "display arenas" to attract mates.

Scuffles and courtship displays made up only fleeting moments in dinosaur lives, of course. Most of the time, discussions of dinosaur social behavior are usually in terms of whether a species lived in social groups, like herds or packs, or relative isolation. The feathery "terrible claw" *Deinonychus* was envisioned early on as a pack hunter, for example, and bone beds composed of hundreds of spiked *Centrosaurus* found in Alberta left little doubt that these horned dinosaurs moved together in massive groups like modern-day bison. But just like the discussion over whether the reptiles were warm-blooded or cold-blooded, our picture of dinosaur social lives does not fit into a neat binary. If anything, a growing body of evidence indicates that many different dinosaur species lived in social groups for part of their lives but later became loners, or perhaps gathered only in unusual circumstances like drought or at an especially abundant food source.

Deinonychus itself, which helped kick off discourse on dinosaur social lives during the height of the Dinosaur Renaissance, offers a case study in how challenging it can be to draw definitive evidence about dinosaur behavior from the fossil record. We want to know how dinosaurs moved and interacted, curiosity that keeps us watching the reptiles race across movie screens and playing with dinosaur toys, but the fossil record is not a book that can be read literally. Paleontologists encounter dinosaurs in their final resting place, which were made sometimes many days, weeks, or months after the dinosaur died—jumbles of disarticulated bones are often from animals that died someplace else and had their remains washed to another spot by localized flooding, and whole dinosaurs, sometimes bloated with decomposition gasses, floated long distances before dropping to the sediment, even out to sea. Working backward from such boneyards to living animals requires not just imagination, but great care.

Even though the bones of multiple *Deinonychus* were found in close association with their possible prey beaked *Tenontosaurus* in the 1960s, and speculation about pack hunting fueled visions of *Jurassic Park*'s raptors in the novel, it wasn't until 1990 that paleontologist John Ostrom formally proposed that the small carnivores worked together to bring down large prey. The quarry that yielded the best array of *Deinonychus* bones contained the remains of at least four individual carnivores, as well as a single subadult *Tenontosaurus*. Shed *Deinonychus* teeth were found across the site, hinting that the carnivores had been feeding when some strange circumstance led to their demise, too. After all, how could multiple *Deinonychus* be buried in the same place with an herbivore if they had not been eating together? And if they were eating together, it wasn't a leap to consider that the dinosaurs had hunted in concert with each other to bring down a bigger meal. Some of the carnivores apparently failed, the larger *Tenontosaurus* seemingly killing several before succumbing itself in what must have been an incredibly violent Cretaceous scene. The idea inspired plenty of paleoart of *Deinonychus* climbing on the larger *Tenontosaurus* and slashing at it with their foot claws, trying to spill viscera and inflict enough trauma to bring the plant-eater down.

Over time, however, the idea that *Deinonychus* were hunting together began to look ever flimsier. The fossil site Ostrom described did not provide any conclusive evidence that *Deinonychus* hunted in a coordinated or even mutually altruistic way. The quarry was a jumble of bones and body parts, showing no direct interaction between the carnivores and the herbivore. It's likely the bodies of the animals had been in close association at the time of death but had become disarticulated and tossed together by the sediment-carrying water that buried them, further muddling the view. Additional sites where *Deinonychus* and *Tenontosaurus* fossils were found together did not show any clear signs of coordinated hunting, either. The pattern just did not match what experts would expect from a group of social carnivores, nor did it line up with how modern lizards, crocodiles, and birds bring down larger animals. Among predatory reptiles like Komodo dragons and Nile crocodiles, in particular, the reptiles are much more antagonistic with each other, even when targeting the same prey. The reptiles sometimes attack and even kill each other in the process of feeding, a scenario that better fits what paleontologists have found at *Deinonychus* quarries. The carnivores were likely drawn to the same place by the prospect of food and fought with each other for their share, creating an ephemeral ecosystem centered around the breakdown of the big herbivore. Even with trackway discoveries that some related *Velociraptor*-like dinosaurs likely traveled together, paleontologists have not yet found any evidence that non-avian dinosaurs engaged in coordinated or hierarchy-based hunting parties like wolves and lions do. The reptiles were probably more like their distant reptile relatives alive today, approaching food sources under uneasy truces that could easily lead to cannibalism as each dinosaur vied for a portion. On top of that, the sites where *Deinonychus* and *Tenontosaurus* were found together are likely unusual associations, representing special circumstances that the carnivores were taking advantage of. Studies of *Deinonychus* claws has shown that they were not slicing bellies of large prey or climbing on the backs of bigger dinosaurs, but pinning down relatively small prey just like redtailed hawks and other birds of prey do. *Deinonychus* was not a megaherbivore hunter, but was more likely to pounce on mammals, smaller dinosaurs, and other creatures that could be held down while *Deinonychus* put their jaws to work.

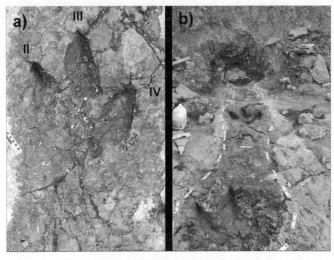

Some of the best evidence for dinosaur social behavior comes from trackways, such as the footprints of several tyrannosaurs that walked side by side found in British Columbia.

Such predatory pressures likely nudged dinosaurs to be social in a different way. Grouping together in herds or similar social groups is one way to reduce the chance that any individual animal is going to be eaten. There are more eyes, noses, and ears to notice the approach of a predator, and, even when an allosaur or other carnivore lunged out of the forest, living in a group meant a smaller chance that any particular dinosaur would be pounced on. Given that all dinosaurs hatched out of eggs and started off life small, even the largest and most imposing of species spent the first years of their lives in danger of being plucked up by a whole array of predators. No surprise, then, that paleontologists keep finding small groups of juvenile dinosaurs.

One such fossil site has been uncovered within Texas's Big Bend National Park. The Late Cretaceous rocks are brimming with bones of multiple *Alamosaurus*, sauropods that could exceed 100 feet (30 m) in length and weigh more than 30 tons (27,000 kg). All the individuals in the Big Bend site, however, are juveniles a fraction of the size. Paleontologists hadn't uncovered a herd made up of babies, juveniles, and adults, as had been proposed on the basis of some sauropod track sites, but rather a collection of juveniles about the same age. Given that sauropod dinosaurs are thought to provide little to no parental care, and the fact the too-big-to-kill adults are often found alone, the emerging picture was of young animals that lived in groups to enhance their chances of survival. If they

The long-necked *Alamosaurus*, seen wandering in the background, formed social groups in their youth to watch out for potential predators.

could grow large enough to be too risky for a tyrannosaur to take a bite, then *Alamosaurus* could begin living on their own and not have to share the lush spots of greenery they fed upon.

The *Alamosaurus* site in Big Bend is not a one-off. Paleontologists have found similar sites of juvenile sauropod dinosaurs in the Late Jurassic rocks of North America—the Mother's Day and Hanksville-Burpee Quarries—in addition to juvenile aggregations of different dinosaur species. *Triceratops* seems to have followed a similar pattern as the sauropods, rare bone beds containing multiple young individuals while more mature *Triceratops* are often found alone. Young theropods have been found in such groups, too. At a fossil site containing fourteen individuals of the ostrich-like dinosaur *Sinornithomimus*, at least eleven were juvenile animals that were likely living together to better watch out for predators.

Multiple horned dinosaur bone beds have been found in western North America. One *Centrosaurus* bone bed contains the bones of thousands of individuals that died in a coastal flood.

Instead of being a life-long behavior, living in groups seems to have been a temporary situation for many dinosaur species with varied evolutionary histories. Keeping close increased chances of growing big and mature enough to strike out solo.

If there's any evidence that some dinosaurs were born into herds and stayed there, it's most likely to be found among the horned dinosaurs of about seventy-five million years ago. In the Late Cretaceous rocks of Alberta and Montana, in particular, paleontologists have found multiple bone beds of horned dinosaurs such as *Centrosaurus*, *Wendiceratops*, and *Medusaceratops*. Each varies in its extent and number of individuals preserved, but the surprising number of multi-individual bone beds that experts continue to turn up seems to be attributable to the consequences of ancient climate and geography.

North America was effectively split in half during most of the Late Cretaceous period. Global temperatures were so warm that there was little to no ice at the poles, meaning sea levels were higher. The increased volume of ocean water spilled over the center of North America, from what's now the Hudson Bay to the Gulf of Mexico, the Western Interior Seaway isolating two subcontinents—Appalachia to the east and Laramidia to the west. The horned dinosaur bone beds are found on the Laramidia side. Despite the fact that the fossil sites are now in landlocked areas, back in the Cretaceous, these places were wet lowlands along the margins of the inland sea. Intense storms would move up these coasts, absolutely drenching the landscape with torrential rains that caused waterways to swell and overflow their banks. With their mouths and noses low to the ground, and bodies not particularly suited to swimming, horned dinosaurs were frequent victims of these storms, literally unable to keep their heads above water. Bone bed 43 in Alberta, perhaps the largest of all, contains the bones from hundreds if not thousands of *Centrosaurus* that all perished in a narrow window, drowned by stormy floodwaters and later picked over by scavenging tyrannosaurs and raptors. While the conditions leading to the formation of smaller bone beds are sometimes debated, the sheer abundance of *Centrosaurus* bones jammed together leaves no doubt that these dinosaurs lived in social groups that included all ages, from juvenile to mature adults. The dinosaurs perished tragically, but because of their unfortunate circumstances, experts now know that some dinosaurs were born into social groups that they stayed with for their entire lives.

Living in a group would have done more than just provided baby stubby-horned *Centrosaurus* protection from the likes of tyrannosaur *Gorgosaurus* and other dangers. Herbivorous dinosaurs, in particular, need to acquire bacteria capable of breaking down plants from their environment. Other *Centrosaurus* could effectively gift their bacteria to the babies by living side by side. The hatchlings would also grow up learning the behaviors of their species from those around them, effectively learning how to be *Centrosaurus* from their neighbors. The dinosaurs didn't lock horns like *Triceratops* did, their skulls lacking the damage of their later cousins, and more likely postured or displayed to get their points across. It was a behavioral setup that worked well for the dinosaurs, seeing them through about a million years of Cretaceous history.

From fights to family dynamics, dinosaurs evolved a wide array of social behaviors that gave rise to different life histories and ways to interact with their environments. The fact that there was no single way to behave like a dinosaur is a testament to just how diverse these reptiles were, developing ways of interacting with members of their own species that relied on everything from the deep, bass-note calls of *Parasaurolophus* to the face-biting of *T. rex*. Had dinosaurs not met with a terrible calamity, they would likely still be here herding and honking together.

CHAPTER 11

Dinosaurs Undone

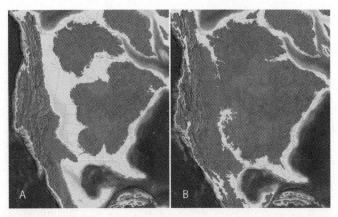

By sixty-six million years ago, the Western Interior Seaway that split North America in two was draining off the continent. A great deal of what we know about the end-Cretaceous extinction comes from swampy coastal lowlands along the edges of the vanishing seaway.

Dinosaurs are practically synonymous with extinction. Our fascination stems from this fact. Dinosaurs are still with us, after all, and most of the time, we hardly pay any attention as they flap or strut by. It's the non-avian dinosaurs, the lost dinosaurs, that have their claws deep in our imagination, creatures that can endlessly fascinate us because none of us will ever see one alive. Not even hypothetical *Jurassic*

Park techno-magic could restore the likes of *Brachiosaurus* and *Triceratops*. Genetic material begins to degrade at death, breaking down so quickly that even under ideal conditions, such as a cool and dry cave, there will be no genetic material left after about six million years. You can't restore what no longer exists.

The longing we feel for dinosaurs is still new, though. In the nineteenth century, when the first scientific concepts of "dinosaur" began to bubble up, there was no great mystery regarding their disappearance. Naturalists thought that the fossil record told a story of progression, and even Darwin's initial formulation of natural selection relied on a sense that improved descendants would supplant their more archaic ancestors. The rock record seemed to bolster this idea, displaying an "Age of Fishes," an "Age of Amphibians," an "Age of Reptiles," an "Age of Mammals," and finally an age of humanity, the supposed culmination of all the previous changes. Dinosaurs represented a temporary state in this tale of ongoing improvement, organisms that were best suited to a warmer, humid, lush world that eventually changed too much for them to last. Geologist and contemporary of Darwin's, Charles Lyell even speculated that if the Earth somehow returned to the conditions in the Jurassic, perhaps creatures like *Iguanodon* and the giant fish-like ichthyosaurs of the sea would return in some form, evolving anew as Earth slipped back into a primordial state.

The fate of the dinosaurs continued to be a non-question for decades more. If anything, the mystery was why they occupied such a long span of prehistoric time. The reptiles were big, strange, and had small brains, all traits assumed to be inferior to the small Mesozoic mammals waiting in the wings. It was puzzling why such awkward creatures should be so numerous and long-lived as a group. Even so, dinosaurs had become such celebrities that scientists and the public alike began to wonder what became of *Brontosaurus* and other reptiles that posed in museum halls. Without a concrete answer, anyone could hazard a guess. Some paleontologists thought dinosaurs were examples of evolutionary senescence. The idea, in short, was that groups of organisms were born, aged, and eventually expired just like individuals do. The fact that the last known dinosaurs were big and often had striking ornamentation like horns or crests was taken as a sign that these animals were putting more of their bodily energy into being big and weird than becoming more intelligent, like mammals did, a sign of a group in decline. The idea carried over into the public sphere. "Going the way of the dinosaur" became a euphemism for big industries or companies too large to rapidly change to shifting markets, and one pacifist group protesting against World War I used "Jingo" the *Stegosaurus* as their icon, an example of putting too much energy into armaments and not enough into intelligence. Almost anyone could take a turn kicking dinosaurs while they were down; reptiles once hailed as majestic upon their discovery now seeming dull and worthy of ridicule.

The paleontologist Michael Benton has described this as the "dilettante phase" of dinosaur extinction ideas. It was clear that dinosaurs, as recognized for most of the twentieth century, suddenly disappeared in the last Cretaceous rocks and had never been found in rocks from the following Cenozoic period, or Age of Mammals. The reason why was a mystery and often raised contradictory possibilities, like Earth's climate becoming too hot or too cold. An ophthalmologist suggested that the horns and odd headgear of some dinosaurs were attempts to shield their eyes from the harsh Mesozoic sun, suggesting cataracts might have been too much for the reptiles to overcome. An entomologist suggested that caterpillars were the culprit. Given that moths and caterpillars predate the Cretaceous extinction, he mused, perhaps caterpillars proliferated too quickly and ate all the available vegetation, leaving no food for dinosaur herbivores, leaving carnivores to scavenge, eat each other, and then perish themselves. A paleontologist raised the possibility of a nearby supernova, as well, the cosmic blast perhaps having deadly consequences on Earth. Hyper-virulent disease, evolutionarily superior mammals, and a bevy of other ideas were all tossed around, but none seemed to fully convince or satisfy. For all the creative ideas, though, no one could offer concrete evidence of why dinosaurs seemed to vanish as abruptly as they appeared.

Now, of course, we know differently. Not all the dinosaurs perished, but almost every lineage perished sixty-six million years ago when a 6-mile-wide (9.6 km) asteroid plowed into Earth's crust. And rather than being the result of a single discovery or realization, the fate of the dinosaurs could be understood only as paleontology itself changed. The key pieces of information didn't even come from the dinosaurs themselves, but from a combination of geological clues and patterns seen among less charismatic creatures like plankton and ocean-dwelling invertebrates that lived at the time.

The mistake that so many dinosaur sleuths repeatedly made was in focusing on the dinosaurs alone. It is not as if *Tyrannosaurus* disappeared and the rest of the world kept turning. The dinosaurs were part of a mass extinction, one that was not fully perceived until computers became more accessible to researchers. In the 1970s and '80s, as paleontologists compiled databases of fossil occurrences, they noticed strange shifts in the record. The story was not of ever-increasing improvement and abundance. There were at least five biodiversity crashes in deep time when many species, from many groups of organisms, became extinct at almost the same time. One of these precipitous drops occurred at the end of the Cretaceous—not detected through dinosaur data, but through the diversity of rapidly reproducing creatures like shelled amoebas called foraminiferans and disc-shaped algae called coccoliths. It was as if the rug was pulled out from under Earth's ecosystems, the demise of the dinosaurs adding to the count. In fact, despite the fact that organisms such as mammals and lizards were thought to be "survivors" of the disaster, these groups also suffered extreme losses around the same time.

The mystery of what happened to the dinosaurs could not be answered by looking at dinosaurs alone. Whatever happened was a swift, global catastrophe that affected life on land as well as in the seas. The leathery-winged, flying pterosaurs vanished, as did the seagoing lizards called mosasaurs. Massive, reef-building clams called rudists disappeared, and the abundant, coil-shelled ammonites went extinct, too. Caterpillars and egg-hungry mammals could not explain the pattern. Whatever happened was a true catastrophe.

It wasn't long after the end-Cretaceous mass extinction was recognized by paleontologists that a new idea for the disaster was proposed. In 1980, in the journal *Science*, a team of geologists proposed that an asteroid, comet, or similar space rock had hit the Earth at the end of the Cretaceous and sparked the cataclysm. The key was iridium, a platinum-group metal rarely found in Earth's crust but commonly found in rocks floating through space. At sites around the world, from Italy to New Zealand, the thin rock layers sandwiched between the end of the Cretaceous and the beginning of the Paleogene periods show extremely elevated iridium levels. While the metal can sometimes be found in molten rock spewed by volcanoes, the relevant rock layers were not volcanic and were dispersed all around the world. Something from space hit the Earth and was so powerfully, thoroughly pulverized that tiny particles were shot back up in a plume that settled all over the world. Especially during the height of Cold War tensions between the United States and the then-USSR, the thought of a giant mushroom cloud blotting out the sun and throwing Earth into a years-long winter was a highly evocative image.

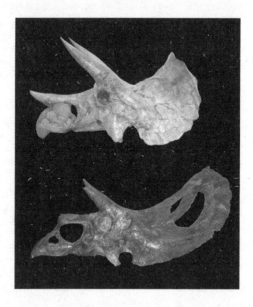

Triceratops and its close relative *Torosaurus* were among the last non-avian dinosaurs alive when the asteroid struck.

Vertebrate paleontologists weren't sold on the new hypothesis. No such event had ever been documented in the fossil record before. The Alvarez Hypothesis, as it was then called, lacked a crater and how the impact would have translated into extinction was entirely unclear. On top of that, paleontologists felt territorial about perceived outsiders suddenly arriving to solve a major puzzle in the history of life without reference to previous work done by dinosaur experts. The Late Cretaceous was a time when the world was cooling, mountain ranges were rising, new connections between continents were opening, and sea levels were dropping. Volcanoes in ancient India, the Deccan Traps, were pouring out incredible amounts of molten rock and spewing greenhouse gases into the air, undoubtedly having an effect on everything from climate to the makeup of the ancient air. Some dinosaur experts also believed that dinosaurs were in decline during their last ten million years. In North America, for example, more dinosaur species were known from rocks about seventy-five million years old, such as those of Alberta's Dinosaur Provincial Park, than in the end-Cretaceous rocks in adjacent areas. No one had found dinosaur fossils in the K/Pg layer, either, and there seemed to be a "three-meter gap" between the last occurrences of dinosaurs like *Triceratops* and the layer at which the asteroid struck. It seemed possible that the asteroid struck when dinosaurs had already vanished, or were barely hanging on, abruptly closing a process of decline that was already in progress.

Decades of investigation and debate continued. Piece by piece, evidence began to click into place for a terrible, unprecedented impact. The infamous impact crater, for example, had been discovered in 1978 when petroleum company geologists were surveying Mexico's Yucatán Peninsula. The divot in the Earth's crust was about 6 miles (9.6 km) across and had formed right at the end of the Cretaceous. Studies of the crater and models of how the impact actually transpired painted a stark picture. The bolide, as it was so often called, must have struck at a low angle and immediately pulverized an incredible amount of rock, creating glass spheres and cracked quartz crystals in the process. The products of the impact were launched high into the air and dispersed, falling all across the ancient Earth. Organisms in the immediate area would have perished instantly, or nearly so, as shockwaves emanated out from the center of impact and towering tsunamis ran outward, powerful enough to leave their mark in the middle of the ancient Pacific. And in a slightly different scenario, perhaps that would have been the worst of it. But geologists and paleontologists have come to realize that the K/Pg impact was a worst-case scenario that affected our planet in unexpected ways.

The debris tossed up from the impact eventually fell back down to the Earth's surface, falling far enough that friction against the air in our atmosphere caused each tiny piece to heat up. The effect of any individual speck was negligible. Together, though, the debris consisted of such volume that the heat created what experts refer to as an infrared pulse. Air temperatures rose to about 500°F (260°C) all around the planet, hot enough to spontaneously ignite dry forest litter. There was no way for dinosaurs, mammals, reptiles, and various other inhabitants of the Mesozoic world to prepare for something so extreme and so fast. Creatures living in the water were given some respite, and small creatures that lived in burrows or could steal them were able to escape the heat

172   THE SHORTEST HISTORY OF THE DINOSAURS

by getting 10 inches (25 cm) or more below the surface of the soil, but there was no escape for *T. rex*, *Alamosaurus*, and so many other living things. Entire ecosystems vanished within the first twenty-four hours of impact, literally scorching the Earth. But that was far from all.

The rocks the asteroid hit were made of limestone, the remnants of ancient reefs that were already fossils in the Cretaceous. As a result, the rock layers were full of sulfur compounds. When aerosolized, as they were by the asteroid strike, the compounds reflect sunlight back out into space and away from Earth's surface. The planet's temperature began to fall, smoke and debris further hindering sunlight's ability to reach plants and other photosynthesizers that provide the basis of many of Earth's ecosystems. In the seas, for example, there was a near-total extinction of photosynthetic coccoliths, while coccolith species that both photosynthesized and fed on organic debris survived. The darkness lasted for three years, and was in fact mitigated by the greenhouse gases produced by the Deccan Traps. The carbon dioxide, methane, and other greenhouse gases released by the prolonged eruptions warmed the climate a few degrees, enough to give living things that survived more of a chance. Even so, about 75 percent of known species rapidly perished. It wasn't just the non-avian dinosaurs that died out. There were also mass extinctions of trees, insects, mammals, reptiles, fish, and more. No part of the tree of life was left untouched by the extinction. The non-avian dinosaurs didn't go extinct because of some peculiar weakness. The reptiles were eradicated practically overnight by circumstances that the survivors navigated through only by luck. So devastating and disruptive were the effects of the impact that even some surviving organisms eventually crumpled long after the worst effects subsided. Ammonites, the coil-shelled cephalopods that had thrived for longer than dinosaurs even existed, fully vanished about one hundred thousand years after the impact. The disaster had so destabilized the seas that ocean ecosystems were not recovering fast enough, perhaps leading the plankton-feeding ammonites to eat their own tiny young and effectively cannibalizing their way to extinction as they hung on in the aftermath.

Most of what we know about the K/Pg disaster comes from the rocks of western North America. In places like Montana, the Dakotas, Utah, Wyoming, Alberta, and more, there are rocks that record the end of the Cretaceous, the K/Pg boundary, and the earliest rocks of the following period, the Paleogene. The before-and-after snapshots allow paleontologists to detect who survived and who perished, sites like those at Corral Bluffs near Denver illustrating how forests grew back after the impact and the ways in which surviving creatures began to change, including mammals. But we still know very little about how the story of extinction and recovery played out in the seas, and in terrestrial habitats elsewhere on the planet. South American paleontologists are making new efforts in Patagonia, and scattered studies have added information from other spots around the globe, but at present we are effectively looking at a planetwide event through a pinhole focused on western North America. Debate and detail will no doubt continue, but the weight of the evidence has left no reasonable doubt that the world of the non-avian dinosaurs was brought to a catastrophic close by a chunk of space rock that had been left over from the formation of our Solar System.

Dinosaurs were not due for such a disaster. The asteroid

was not closing a story already in the middle of its conclusion. While it is true that we know more dinosaur species from seventy-five-million-year-old rocks than those of sixty-six million years ago, a large part of the reason why is because environments amenable to preservation were disappearing when the asteroid hit. In North America, for example, the Western Interior Seaway that created vast tracts of swampy lowlands good for preserving dinosaurs was disappearing, meaning that there were far fewer places where dinosaurs could have been preserved as fossils. Had the seaway remained, experts would likely find a whole new array of dinosaurs on the continent, from ancient Alaska down to Mexico. Such as it is, it

Even if small non-avian dinosaurs like *Acheroraptor* survived the first twenty-four hours after impact, the stresses of the post-impact winter left beaked birds as the only surviving dinosaurs.

seems conditions sixty-six million years ago weren't right to fold as many dinosaurs from such a broad area into the fossil record, an invisible part of the story that nevertheless has to be accounted for in considerations of how the extinction played out. The pieces of the fossil record we've found hint at all the parts we can never directly perceive, instead we are fumbling around at the edges.

Of course, one group of dinosaurs survived the devastation. Birds are our living dinosaurs. Their survival is a mystery within the broader puzzle of the K/Pg extinction. What allowed the ancestors of pigeons and penguins to survive, but not relatives of *Triceratops* or even small raptors that were small enough to find refuge during the first terrible day after impact?

From the Late Triassic period onward, dinosaurs filled the Mesozoic world in forms from tiny to huge. The end of the Cretaceous was no different. Even though imposing dinosaurs such as horned *Torosaurus* and, of course, *T. rex* steal the show, small birdlike species lived right alongside the giants. Many of them have gone unnoticed until recently, not just for their rarity, but because historic archeological expeditions were mostly looking for impressively big museum specimens. The small dromaeosaur *Acheroraptor*, a turkey-size feathered raptor that lived in the same Hell Creek Formation habitats as *Triceratops*, was named only in 2013. Nevertheless, among these smaller finds, paleontologists have uncovered birds with beaks, toothed birds, and small, feathery raptors that were all living alongside each other as the asteroid struck.

Body size was not the deciding factor. Beaked birds, toothed birds, and raptors overlapped in their size ranges, and many were undoubtedly small enough to make use of burrows—their own or created by other animals—during the first, fiery twenty-four hours after impact. In fact, it's possible that members of all three groups lived long enough to make it into the gloomy impact winter that followed. Ultimately, though, it was only beaked birds that survived to thrive and diversify from those early Paleogene days until today. Diet might have made all the difference. In a 2016 study of Late Cretaceous birds and birdlike dinosaurs, paleontologist Derek Larson and colleagues proposed that beaked birds were pre-adapted to survive on what little food remained in an ashen world post-impact. Both small raptors and toothed birds ate insects, lizards, and smaller animals. While some of these morsels surely survived, their populations would have been much, much smaller than they were before. Forests that once hosted seemingly endless insects, for example, were entirely wiped out, leaving any survivors far less to feed upon. Beaked birds, however, had long ago shifted to a more varied diet that included much more plant material than their toothy coun-terparts. With adaptations like gizzards and gastric mills to break down tough plants that they swallowed whole, beaked birds were able to subsist on seeds, nuts, and other resilient plant parts that remained in the soil. The dirt insulated a sig-nificant amount of life, enough for temporary ecosystems to form and set up life to grow back throughout the Paleogene. In fact, recent research has found that fungus-farming ants evolved their relationship with their food source in the direct aftermath of the K/Pg impact, underscoring how surviving species leaned on each other as they formed ecosystems unlike any seen before. And dinosaurs, in their most feathered forms, would be part of it.

It's strange to think that a group of animals as long-lived and varied as the dinosaurs left only one subset of one group behind. To say birds survived is too simple. Entire groups of birds, including all toothed birds, died out. Dinosaurs came within a feather's breadth of entirely vanishing after more than 160 million years of existence. If anything, their near obliteration speaks to how harsh Earth's fifth mass extinction truly was. Even though other mass extinctions were greater in severity, or dragged on for longer amounts of time, the end-Cretaceous catastrophe was entirely unprecedented, introducing circumstances that life on Earth had never endured before. This was not another volcano-driven disaster that turned the climate hot, acidified the seas, and altered the global carbon cycle (though those are bad enough). The consequences of the K/Pg impact were unlike any terrestrial trigger, both severe and strange at once. The event is a reminder that life on Earth is not only shaped by the biotic and abiotic processes inherent to our planet, from who eats whom to continental drifts, but that our planet is part of a constantly-shifting solar system,

The earliest primate, *Purgatorius*, survived the K/Pg extinction.

something as seemingly distant as astrogeology still impacting and consequently shaping our world.

The only reason that we are here to discuss these events is because of this extinction and the survival of our earliest primate ancestors. The oldest primate, the shrewlike *Purgatorius*, lived in western North America when the asteroid hit. *Tyrannosaurus* vanished. *Purgatorius* survived. And in a world devoid of dinosaurs save for birds, mammals rapidly spread through ancient forests that could finally grow thick. Early primates quickly diversified and began to evolve key features such as forward-facing eyes and grasping hands, traits they may never have acquired if Paleogene forests had not been able to form multilayered canopies full of fruit and crunchy insects. If the deadly asteroid had missed Earth, hit a different place on the planet, come in slower, or at a different angle, then the mass extinction might have been canceled or, at the very least, significantly altered. It took circumstances that life on Earth had never experienced to wipe out the non-avian dinosaurs, and there is every reason to believe the reptiles would have continued blossoming for millions of years to come if they had been given a reprieve. Even though dinosaurs were not directly suppressing mammals, they still shaped the world in such a way that the origin of tree-dwelling apes that someday ventured to the ground would have been lost or so altered that humans wouldn't come to be. The effects of the K/Pg impact are still felt today, down to our very existence and ability to ponder what became of our favorite dinosaurs. It's a feeling of subtle and constant grief all dinosaur lovers carry, that if the great reptiles had lived then we would not be here to miss them.

CHAPTER 12

How to Become a Fossil

"Dinosaur" is almost a magic word. All we need to do is say it and we begin envisioning prehistoric landscapes filled with strange and ancient reptiles, horns and teeth gleaming in the Mesozoic sunshine. The amazing reptiles are creatures of a lost world, separated from us by a terrible extinction and more than sixty-six million years. And yet dinosaurs are still with us, and not just as birds. Bones, tracks, skin impressions, and even fossilized feces are all dinosaur remnants that have crossed the span of time between us and the Mesozoic era—transformed but still tangible.

Our knowledge of the dinosaur fossil record is growing each day. Dozens of new species are described by experts each year, not to mention new specimens of already-known dinosaurs. Many new dinosaur species have already been found, prepared, and studied, merely waiting for the scientific publication process to introduce them to us. But our knowledge of the dinosaur fossil record is not only incomplete, but always will be. Fossilization is not the norm, paleontologists well know, but is a rare process that relies on special circumstances. It's truly a wonder that we have any fossils at all.

Naturalists have been grappling with the irregular and fortuitous nature of the fossil record since the nineteenth century. Drawing inspiration from his friend and colleague Charles Lyell, Charles Darwin devoted an entire chapter of *On the Origin of Species* to "the imperfection of the fossil record" to explain why the transitional species his evolutionary theory predicted had not yet been identified. If we think of Earth's rocks as a great geological book, Darwin wrote, the tome would be missing whole chapters, pages, paragraphs, sentences, and words. Earth is not some great geological onion with continuous layers recording what transpired over the planet's extent, but instead represents places where sediment was laid down in a discontinuous series. Part of the reason western North America is a focal point for studies about the fate of *T. rex* and *Triceratops* is because rocks in the Dakotas, Montana, and Alberta represent both the end of the Cretaceous and the beginning of the following Paleogene period, providing before-and-after snapshots of what life was like between sixty-six million and sixty-five million years ago. In other parts of the world, however, half of the picture is missing, preserving only the Cretaceous or Paleogene but not both. The fossil record is, by its very nature, incomplete. Not all environments can create fossils, creating a patchwork of rock layers that paleontologists are constantly comparing against each other to perceive the big-picture evolutionary and ecological changes that have played out on our planet. We will always have more fossils from along ancient streams and the bottoms of ancient lakes than we will from upland forests or rocky deserts where exposed stone is being worn back down into sand. And even within habitats where enough mud, sand, ash, or other sediment was being deposited to preserve fossils, the rocks contain only a fraction of the living things that once lived in those environments. The chances are generally against an organism or its traces being preserved as a fossil, a fact paleontologists know well through a subdiscipline called taphonomy. This science is generally called "what happens between death and discovery," a way to look at the available clues and work out the circumstances of a fossil's backstory. Taphonomy helps explain why some fossils are found only as broken-down fragments while others are complete, articulated skeletons, and also why the discovery of any dinosaur fossil is a victory against the odds.

Complete, articulated skeletons of dinosaurs, such as this sauropod *Spinophorosaurus*, are rare finds.

Taphonomy is practically as old as paleontology itself. Fossil discoveries of the nineteenth century often inspired the question of how such strange remains came to reside where they were found and what they might indicate about how the world changed. Of course, dinosaurs and other prehistoric reptiles were a bit too unusual compared to modern species to fully comprehend their lives and fossil afterlives. Fossilized mammals were more approachable, like the Ice Age hyenas that resided in England's Kirkdale Cave. William Buckland, of *Megalosaurus* fame, was fascinated by the collection of chewed bones found inside the cave, which must have served as a den for ancient spotted hyenas. Buckland compared the damage on the fossil bones and fossil dung found in the cave with the gnawing and scat of living hyenas in menageries, confirming that the cave really had been the site of many ancient hyena meals. Studies such as Buckland's began a fossil subdiscipline concerned with working out why and how prehistoric creatures came to be in the state we find them, from how carcasses decompose to why dinosaurs often take on a death pose with the head thrown back and tail raised.

Dinosaur taphonomy is its own unique subset, influenced not only by geological details, but the unique ecology of the distant past. Regarding dinosaur body size, for example, there appears to be a sweet spot for better preservation chances, animals that are not too small to be easily destroyed by decomposition and not too large to require unusual events to bury them. Places where dinosaurs could be preserved mattered, too; dinosaurs would somehow have to find their way into habitats where sediment was being laid down rather than swept away. At any given time in the whole Mesozoic, the number of habitats capable of preserving dinosaurs was limited. Even preservation by unusual means, such as dinosaur feathers encapsulated by tree resin, still requires burial and sediment that eventually becomes stone. And within these places with fossil potential, most living things will either be eaten by others or broken down by bacteria before there's even a chance of burial. Paleontologists have suggested that hatchling and juvenile dinosaur specimens are so rare, for example, because the naïve reptiles were easy prey for the large carnivores that stalked prehistoric habitats, meaning that baby dinosaurs would have to perish in rare events like local floods or dune collapses to be preserved as fossils. And even when dinosaurs grew to immense sizes and survived to old age, they still would have to die in a place where sediment was being laid down, so they could be covered up before carrion-eating dinosaurs, birds, beetles, and bacteria broke down their bodies. All of the contenders for the largest dinosaur species of all time are known only from incomplete skeletons that began to fall apart before some of the bones could be covered up. We can see the same forces at work today. If you spot the bones of an animal during a walk in the forest or along a beach, chances are that the carcass has already been scattered, broken down, and is on its way to total obliteration rather than being folded into sediment that will allow those parts to become fossils.

And yet paleontologists have amassed an impressive fossil menagerie, with new fossil finds made every year. While the fossil record is perhaps not as generous as we might wish it to be, we nevertheless know far more about prehistoric life than early geologists ever could have dreamed. The right circumstances occurred just often enough, through hundreds of millions of years, to let us at least outline the lives of dinosaurs and their neighbors throughout the Mesozoic.

No two dinosaur skeletons are the same. The afterlife of each and every one is a little different. The reasons a dinosaur perished, what kind of habitat they died in, how long it took for them to be buried, the nature of the sediment that covered them, the minerals of the surrounding rocks, and even the amount of local rainfall all mattered for what dinosaur skeletons became. In many cases, dinosaurs partially decomposed or even had all their flesh rot away before burial, streams jumping their banks or similar circumstances gathered the loose bones and deposited them together in a jumble. The famous Carnegie Quarry of Dinosaur National Monument represents such a scenario—dinosaurs that died during Jurassic dry seasons gathered up by rushing water when wet season rains returned, carrying the bodies, body parts, and bones to the same spot, the larger carcasses creating a kind of logjam that caused other dinosaur carcasses and body pieces to pile up. Then again, experts have found fossils like *Mei long* from China, a tiny theropod dinosaur that was buried in fine-grained ash as it slept in a birdlike position, nose tucked under an arm. The dinosaur perished and was buried as it was at rest, the skeleton not only complete but articulated with each bone more or less in life position. Paleontologists have found dinosaur bodies with no skulls, skulls with no bodies, isolated teeth, bone beds of hundreds of

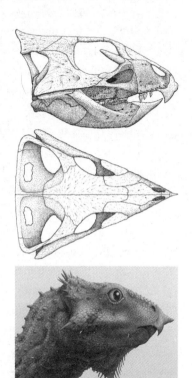

Small dinosaurs such as the horned *Aquilops* were more likely to be eaten, decayed, or otherwise destroyed before preservation, making them harder to find. So far, only the skull of one *Aquilops* is known.

individual dinosaurs accumulated over years, and much more, every single discovery presenting its own unique puzzle.

Burial in some kind of sediment, of course, is the common factor in all these outcomes. Whether in sand, mud, silt, ash, or some other residue, dinosaur bodies and tracks had to be covered up and protected from the elements to have a chance of becoming fossils. (And even then, some dinosaur bodies certainly decomposed entirely before they could be mineralized and survive for millions of years enclosed in the geologic record.) Once covered in sediment, water from rain, streams, or other sources could percolate down through the gaps in the encasing layers, carrying dissolved minerals within it. Bones, being a very spongy tissue made of many microscopic cells, soaked up the mineral-laden water, the minerals beginning to replace the bone tissue at a microscopic level. At this point, however, it is important to stress that fossil bones have not "turned to stone," as is often said. Dinosaur bones are replaced by minerals to differing degrees, sometimes across elements in the same skeleton. Sometimes the mineralization is a complete replacement, but more often there remains some original, degraded tissue inside the bones. Broken-down collagen and other biomolecules have been found inside dinosaur bones and even in preserved skin, no longer appearing like fresh tissue but still containing traces of the original biological material. Paleontologists have only recently realized that such original biochemical traces of non-avian dinosaurs exist, so much so that the discipline of molecular paleontology is only just beginning. Even though accessing dinosaur DNA is impossible after so long, paleontologists are still finding an array of biological remnants in dinosaur bones that have not been completely replaced by minerals.

But we don't even have to get that deep into the bones to see that the way dinosaurs became fossils was so varied it was practically individualized. Dinosaur skeletons are all sorts of different colors, for example, because of the minerals being carried in the water that percolated through the sediment and slipped inside the bones during the fossilization process. Different minerals create different colors, and when the fossil bones are exposed to the elements, in a process called weathering, those shades can change again. A dinosaur toe bone found exposed on the surface might be white with touches of purple, for example, but the rest of the skeleton in the rock might be rust red as it's been encased by the rock. Dinosaur bones can even be opalized, such as some fossils found in the area of Australia's Lightning Ridge. Each dinosaur bone and skeleton is a kind of time capsule of the place the animal was buried, transformed, and withstood the test of time in the rock record. The process did not take sixty-six million years or more. Paleontologists have found plenty of mineralized fossils from animals that lived a million years ago, and even younger, meaning that even during the time of *T. rex* there was an extensive dinosaur fossil record spanning hundreds of millions of years. Imagine the geologic upheavals of the Late Cretaceous, for example, when mountain ranges were bringing older rocks to the surface. It's entirely possible that tyrannosaurs, ceratopsians, ankylosaurs, and their neighbors walked right by the eroding fossils of their Triassic relatives that were preserved millions of years before, pressed into the Earth's crust, and brought back to the surface millions of years before the earliest humans would even exist.

"Dakota" the *Edmontosaurus* demonstrates how lying exposed may have helped dinosaur skin and its impressions be preserved.

As paleontologists learn more, they've realized that the fossilization process is relatively rare but can happen in so many ways that even partially decomposed dinosaurs could still make fantastic fossils. The fate of a particular duck-billed dinosaur nicknamed "Dakota" is an instructive case. The sixty-seven-million-year-old *Edmontosaurus* skeleton is covered in broad swaths of preserved skin, including a right front leg that indicates hadrosaurs had something like a hoof covering their mitt-like front feet. Paleontologists have been uncovering and studying such fossils since 1909, when professional fossil hunter Charles H. Sternberg found a different *Edmontosaurus*—then called *Trachodon*—wrapped in skin impressions. Experts were certain that such fossils required what's often called exceptional preservation, circumstances that seem rare even compared to the slim chances of more usual forms of fossilization. Dead dinosaurs that stayed relatively intact were thought to be quickly covered by sediment from flooding rivers or other events that could move enough sand and mud, the blankets of tiny grains creating impressions of the skin before it decayed away. But the story of Dakota reveals a different pathway for dinosaurs with their squishy parts still represented in the rock. In some ways, the *Edmontosaurus* had to start rotting away for there to be a chance for its soft tissues to become fossils.

We'll probably never know what killed Dakota, but the dinosaur's afterlife can still be drawn from the fossil. A 2022 study by paleontologist Stephanie Drumheller and colleagues found evidence that the dinosaur's skin shows signs of scavenging and drying out. Instead of being buried at the time of death and soon after, the *Edmontosaurus* had been exposed to the sun and elements for weeks to months after death. The carcass was not dismantled by adult tyrannosaurs, but instead had been opened up and nibbled on by smaller carnivores, perhaps a Cretaceous crocodile. As a result, the decomposition gases, body fluids, and microbes inside the dinosaur escaped, letting the viscera and muscle inside rot away as the skin dried and clung to the skeleton. The dinosaur's tough skin was able to withstand what the softer parts could not, and so when some local event sent sediment cascading over the dinosaur's body to bury it, the shrink-wrapped skin was preserved alongside the bones. The origin of the detailed dinosaur fossil did not require anything out of the ordinary, following decomposition pathways commonly seen among living animals. And given the number of hadrosaur fossils found with skin and skin impressions, it seems what was once thought of as unusual was actually relatively common in the dinosaur fossil record. There are probably more dinosaur "mummies" and specimens with preserved soft tissue clues than anyone expected. Drumheller and colleagues concluded, the fact that

Dakota was exposed for a long time and still is one of the best *Edmontosaurus* specimens known "suggests that such types of evidence may be substantially more common than previously assumed."

Even fossilization in amber is a more complex process than it might seem. Thanks in no small part to the *Jurassic Park* franchise, prehistoric creatures enclosed in ancient tree resin often seem pristine and entirely intact. The fact of the matter is that the insects, birds, reptiles, dinosaur body parts, and other organisms encased in amber are just as transformed as the bones in the rock. The process goes something like this. Let's say a small dinosaur dies near the base of a prehistoric conifer. Many conifers, which were abundant during the Mesozoic, produce resin when they're scratched or otherwise harmed. The yellow, sticky substance is a kind of natural bandage, helping to close up places in the bark and wood where insects, fungus, bacteria, or other potentially harmful organisms could get into the heart of the tree. So, the tree the small, fuzzy dinosaur died beneath has such a gash and drips resin onto the body, enclosing it in the syrupy substance that soon dries. The dinosaur is now encased in tough outer layers that will help prevent scavengers from breaking the carcass apart and bacteria from breaking it down.

Now resilient to the elements and other destructive forces, the amber-encased dinosaur body is itself transported by water and buried just like a dinosaur bone would be. Covered up, it's folded into the sediment. And as water filters through that sediment, it also gets inside the resin casing through microcracks. The soft tissues of the dinosaur's body—its feathers and scales and other parts—are replaced by minerals just as a bone would be, creating an intricate, mineralized replica of the dinosaur. Uncovered more than sixty-six million years later, such fossils give us the shape of ancient life in a way we could never expect to see, but they are still transformed entities that are not simply perfectly preserved. In a way, tree resin is another form of sediment just like sand, substances that gave dinosaurs a chance to be transformed enough to survive millions of years before discovery. The fact that resin-producing trees spread over wide swaths of the planet between 125 and 75 million years ago, a time called the Cretaceous Resinous Interval, helped create the fossil record we know today, another way in which environmental happenstance shaped what we know about the past.

The armored dinosaur *Borealopelta* is so well-preserved that paleontologists have been able to work out the dinosaur's color from microscopic clues in its skin and armor. In life, the dinosaur would have had a reddish upper and lighter lower body.

Naturally, the dinosaur fossil record is not made up of only body fossils. Trace fossils—preserved steps and recorded activity—are an important part of the story, especially because traces are effectively fossil behavior. Bite marks can indicate who was eating whom, and something as simple as a dinosaur trackway can reflect a dinosaur's height, how fast they were moving, and perhaps even what they were traveling after. In some cases, the persistence of these fleeting moments in the fossil record is relatively easy to understand. Toothmarks on a bone will be recorded as the bone itself is covered and mineralized. Footprints are a different story, even if the principles are similar to body fossils. A series of footsteps made by a limping carnivore about 150 million years ago, now preserved at a trail called Copper Ridge in eastern Utah, offers an example.

Unless a dinosaur literally dies in its tracks, the exact identity of a trackmaker is often impossible to say for sure. Paleontologists usually narrow down the possible candidates by what dinosaurs lived at the same time and how the shape of the track matches fossil feet. Huge three-toed tracks with sharp toe claws found in sixty-eight- to sixty-six-million-year-old rock of western North America, for example, almost certainly were made by *Tyrannosaurus* because no other carnivorous dinosaur of that size is known from that time and place. But sometimes tracks and body fossils are found in different layers, or don't coincide, and so experts focus on narrowing down the general type of dinosaur that likely left the footprint, such as "stegosaur" or "sauropod." In the case of the Copper Ridge tracks, the footprints were most likely made by an *Allosaurus*, the most common carnivore in the Morrison Formation, but the three-toed tracks could have just as easily been made by rarer predatory species such as

Ceratosaurus or *Torvosaurus*. All experts can be sure of is that it was a carnivorous, sharp-clawed species that made the tracks, the distance between the right and left tracks indicating that the dinosaur was moving with a limp. Sometime around 150 million years ago, perhaps by the muddy edge of a Jurassic lake, an *Allosaurus* that had a bad run-in with an *Apatosaurus* just a bit too big to be safely hunted walked with a hitch in its step, creating a series of impressions in the sandy sediment.

The tracks were not disturbed by other animals. They were pressed in so firmly that they kept their shape, perhaps even drying a little in the sunlight and becoming a little harder. And as wind caused the lake water to lap, moving bottom sediment with it, sand covered the tracks, not wiping them away but filling them in. The track acts as a natural mold, the infilling sediment creating a cast, or counterpart, as it settles and hardens. As those layers get covered by even more sediment, the layers are pressed deeper and preserved, much later pushed up above the surface again. As erosion worked away the overlying layers, the preserved tracks were exposed, the cast destroyed but the mold, the footprint, remaining.

Paleontologists might find the actual tracks, their natural casts, or both. And many dinosaurs were so large that their footprints made what experts call "undertracks," distortions in the sediment that are like fuzzy versions of the footprint. The behavior of the animal and the nature of the sediment, such as grain size and wetness, all affect the resulting track, creating a record of footprints ranging from vague potholes to those so delicately preserved that you can see individual scales on them. Each says something about what dinosaurs were doing and where they lived, from forests to sandy coastlines. Where bones tell us what dinosaurs could do, tracks and other traces tell us what dinosaurs actually did.

The dinosaur fossil record is continuing to grow. Birds, as we well know, are living dinosaurs. Their fossil record continues through the Paleogene to today. Whether we think of the species as common or exotic, bird footprints, feathers, bones, and more are all being folded into the fossil record. A million years from now, they'll speak to what our world was like. Fossils are not just things of the past, but are even now forming all around the world.

Afterword

Most dinosaurs never made it into the fossil record. What we know today is a tiny—but amazing—fraction of all the dinosaurs that ever lived.

If there is anything that we can say for sure about dinosaurs, it is that they'll keep changing. In the past two hundred years, our favorite prehistoric creatures have stomped through our imaginations as immense lizards, leaping oddities, tail-dragging dullards, and rainbow-colored fluffballs. We know dinosaurs better than ever before, but every new discovery raises new questions. It couldn't be any other way. Based upon the number of dinosaurs known so far, and the amount of potentially dinosaur-bearing rock formations around the world, we have discovered only about a third of the dinosaurs that are likely out there. We are trying to understand dinosaurs before we have even found them all, and the vast majority of known species will significantly change if we find better skeletons. Understanding dinosaurs will always require reconciling what little we've found with what we don't know.

But it would be foolish to say that any given dinosaur mystery will be unsolvable, or that we will never know for sure. For decades prior to now, any discussion of dinosaurs wasn't complete without the sentiment that no one knew what happened to them and that we likely never would. Now we do: a catastrophe so unprecedented that it's surprising even a quarter of Earth's species managed to survive the fire and cold. The colors of dinosaurs, too, have traditionally been treated as an area that we couldn't possibly gain insight into, giving paleoartists and kids with crayons alike free license to color dinosaurs any way they pleased. Who could say whether or not *T. rex* was Malibu pink? We can still exercise that imagination, but paleontologists have also found ways to start filling in dinosaur shades, finding that they shared many hues and patterns with modern birds. Paleontology itself has put decades of effort into changing dinosaurs from monstrous oddities to once-living animals that were part of complex ecosystems, organisms that we can approach and attempt to understand just like living species.

Science is not a matter of placing facts on a shelf, however. Fact and theory are constantly interacting with each other, and an unexpected new discovery can always cause a vague possibility to snap into focus. For as much as they did not yet understand dinosaurs, nineteenth-century paleontologists were beginning to understand just how much of the fossil record was unknown and that petrified surprises would undoubtedly astonish future experts. In the face of such incomplete knowledge, imagination is essential.

Science and imagination are not opposed. If anything, science thrives when experts consider what might have been possible and allow themselves to bring logic to what might seem to be fanciful or unreasonable possibilities. The idea that birds might be living dinosaurs did not arise just because solid, irrefutable evidence was available. The notion started as speculation, a possibility based on passing resemblance. The idea stuck in the minds of experts as they continued to search for evidence, tested by the fossil record itself. Our modern view of dinosaur "weapons" as ornaments and socially influenced structures, too, required that paleontologists stop assuming that what looked like spears and shields were weapons and opened up the possibility that strange dinosaur features stood out to the animals as much as they entrance us. A hypothesis is a formalized way of saying "What if?," a starting point that sets experts off into deserts and poring over collections of bones to put the question through a logical trial. Our beloved, changing dinosaurs are a result of a healthy scientific process, changeable creatures that are envisioned thanks to the efforts of generation after generation of people who can't help but be fascinated by the bones of dear departed reptiles.

As much as we might be frustrated by the imperfection of the fossil record, with every dinosaur paleontologist heading into the field wishing for complete skeletons and remnants of soft tissue, it's the gaps in the fossil record that allows us to understand dinosaurs as we do. If we had a totally complete record of every dinosaur that ever lived, it would be incredibly difficult to draw distinctions between species and to follow the paths of their evolution. The fact that we get only glimpses, coming to us piecemeal and with great effort, provides spaces where we can work from broad outlines toward detail. And especially as we are fascinated by the strange nature of dinosaurs, our frustration at fossil gaps helps to generate new ideas about how dinosaurs are related, how they lived, and what they looked like. Uncertainty is a significant driver of our imagination, and nature often exceeds even our wildest dreams.

Initially known from arm and shoulder bones, *Deinocheirus*'s true form was a mystery for decades. When the rest of the body was found, the dinosaur was stranger than paleontologists expected.

The story of a mystery dinosaur underscores the wonder we can expect in the years and decades to come. In 1970, paleontologists Zofia Kielan-Jaworowska and Ewa Roniewicz named an enormous dinosaur from little more than a pair of 8-foot-long (2.4 m) arms. Named *Deinocheirus*, the "terrible hand," the dinosaur was a huge theropod. Little more could be said about it. Years went by without any new significant material. Some experts thought that the dinosaur was one of the "ostrich mimic" dinosaurs like *Struthiomimus*, but much bigger than had ever been found before. There wasn't much more to find from the original fossil site, and additional skeletons remained elusive. But in 2013, at the annual Society of Vertebrate Paleontology meeting held in Los Angeles, California, that year, paleontologist Yuong-Nam Lee stood before the assembled crowd to announce that two additional *Deinocheirus* skeletons had been found. Despite damage by poachers and some of the bones temporarily making their way into the private fossil market, Lee and coauthors were able to bring all the known material together to finally reveal what the mystery dinosaur looked like. Attendees gasped when the new restoration was presented on the screen. *Deinocheirus* was not just like a big ostrich or any of its close relatives. The dinosaur was about as large as *T. rex*, but had a long, beaky snout, a low sail on its back, and a tail tipped in prominent feathers. *Deinocheirus* was an ornithomimosaur, but one that somehow mimicked duck-billed hadrosaurs, the sail-backed spinosaurs, and its relatives all in one animal, an omnivore that walked the swampy habitats of Mongolia about seventy million years ago. Of all the speculative illustrations of the animal done over the years, none had even come close to the true form of *Deinocheirus*.

Even familiar dinosaurs have undergone makeovers ranging from slight to significant. The small horned dinosaur *Psittacosaurus* was named in 1923 but the bristles on the dinosaur's tail were unknown until 2002. A fossil described in 2013 revealed that the shovel-beaked *Edmontosaurus*, despite already being known from several skeletons preserved with soft tissue, had a rounded "cock's comb" on its head that no one had ever seen before.

Such finds will no doubt continue, introducing paleontologists to strange new forms and modifying what we thought we knew. And even as you have learned more about dinosaurs through these pages, hold your image of dinosaurs gently, as it is undoubtedly going to keep changing, part of a centuries-old relationship fueled by our fascination. Even tomorrow, we might wake up to find our favorite dinosaurs are a little different from what we thought when we went to sleep. Change is the very nature of the dinosaur.

Further Reading

Introduction

Kim, H. J., Paik, I. S., Lim, J.-D., "Dinosaur track-bearing deposits at petroglyphs of Bangudae Terrace in Daegokcheon Stream, Ulju (National Treasure No. 285): Occurrences, paleoenvironments, and significance in natural history," *Korean Journal of Heritage: History and Science* 47, no. 2 (2014): 46–67.

Nesbitt, S. J., Desojo, J. B., and Irmis, R., "Anatomy, phylogeny, and palaeobiology of early archosaurs and their kin," Geological Society London, Special Publications, 379 (2013): 1–7.

Nieuwland, I., *An American Dinosaur Abroad* (University of Pittsburgh Press, 2019), 21–48.

Percival, L. M. E., Ruhl, M., Hesselbo, S. P., et al., "Mercury evidence for pulsed volcanism during the end-Triassic mass extinction," PNAS 114, 30, (2017): 7929–34.

Simpson, M., "The walk that changed history: new evidence about the discovery of the Iguanodon," *Deposits*, 2020, depositsmag.com/2020/07/17/walk-that-changed-history-new-evidence-about-the-discovery-of-the-iguanodon.

Staker, A. R., "The earliest known dinosaur trackers of Zion National Park, Utah," in *The Triassic-Jurassic Terrestrial Transition, New Mexico Museum of Natural History and Science Bulletin* 37, ed. Harris et al., New Mexico Museum of Natural History and Science (2006): 137–39.

Troiano, L. P., dos Santos, H. B., Aureliano, T., Ghilardi, A. M., "A remarkable assemblage of petroglyphs and dinosaur footprints in Northeast Brazil," *Scientific Reports* 14, no. 1, (2024): 6528.

Chapter 1: How to Make a Terrible Lizard

Griffin, C., Wynd, B., Munyikea, D., Broderick, T., Zondo, M., Tolan, S., et al., "Africa's oldest dinosaurs reveal early suppression of dinosaur distribution," *Nature* 609 (2022): 313–19.

Kammerer, C. F., Nesbitt, S. J., Flyn, J. J., Ranivoharimanana, L., and Wyss, A. R., "A tiny ornithodiran archosaur from the Triassic of Madagascar and the role of miniaturization in dinosaur and pterosaur ancestry," *PNAS* 117, no. 30 (2020): 17932–36.

Labandeira, C., "The fossil record of insect extinction: new approaches and future directions," *American Entomologist* 51, no. 1 (2005): 14–29.

Metcalfe, I., Crowley, J., Nicoll, R., Schmitz, M. "High-precision U-Pb CA-TIMS calibration of Middle Permian to Lower Triassic sequences, mass extinction and extreme climate-change in eastern Australian Gondwana," *Gondwana Research* 28, no. 1 (2015): 61–81.

Müller, R., Garcia, M., "Oldest dinosauromorph from South America and the early radiation of dinosaur precursors in Gondwana," *Gondwana Research* 107 (2022): 42–48.

Nesbitt, S., Barrett, P., Werning, S., Sidor, C., Charig, A., "The oldest dinosaur? A Middle Triassic dinosauriform from Tanzania," *Biology Letters* 9, no. 1 (2013).

Nesbitt, S., Langer, M., Ezcurra, M., "The anatomy of *Asilisaurus kongwe*, a dinosauriform from the Lifua Member of the Manda Beds (~Middle Triassic) of Africa," *The Anatomical Record* 303, no. 4 (2019): 813–73.

Olsen, P., Sha, J., Fang, Y., et al., "Arctic ice and the ecological rise of the dinosaurs," *Science* 8, no. 26 (2022).

Chapter 2: Eggs and Nests

France," in *Dinosaur Eggs and Babies*, ed. Carpenter, K., Hirsch, K. F., and Horner, J. R. (Cambridge University Press, 1994), 31–34.

Erickson, G. M., Zelenitsky, D. K., Kay, D. I., and Norell, M. A., "Dinosaur incubation periods directly determined from growth-line counts in embryonic teeth show reptilian-grade development," *PNAS* 114, no. 3 (2017): 540–45.

Ezcurra, M. D., Butler, R. J., "The rise of the ruling reptiles and ecosystem recovery from the Permo-Triassic mass extinction," *Proceedings of the Royal Society B* 285 (2018).

Grellet-Tinner, F., Fiorelli, L. E., "A new Argentinean nesting site showing neosauropod dinosaur reproduction in a Cretaceous hydrothermal environment," *Nature Communications* 1, no. 32 (2010).

Hirsch, K. F., Stadtman, K. L., Miller, W. E., and Madsen, J. H., "Upper Jurassic dinosaur egg from Utah," *Science* 243, no. 4899 (1989): 1711–13.

Kirkland, J., "Fruita paleontological area (Upper Jurassic, Morrison Formation), western Colorado: an example of terrestrial taphofacies analysis," *New Mexico Museum of Natural History and Science Bulletin* 36 (2006): 60–95.

Kundrát, M., Coria, R. A., Manning, T. W., et al., "Specialized craniofacial anatomy of a titanosaurian embryo from Argentina," *Current Biology* 30, no. 21 (2020): P4263–69.E2.

Lee, A. H., and Werning, S., "Sexual maturity in growing dinosaurs does not fit reptilian growth models," *PNAS* 105, no. 2 (2008): 582–87.

Reisz, R. R., Evans, D. C., Roberts, E. M., Sues, H.-D., and Yates, A. M., "Oldest known dinosaurian nesting site and reproductive biology of the Early Jurassic sauropodomorph Massospondylus," *PNAS* 109, no. 7 (2012): 2428–33.

Schweitzer, M. H., Zheng, W., Zanno, L., Werning, S., Sugiyama, T., "Chemistry supports the identification of gender-specific reproductive tissue in Tyrannosaurus rex," *Scientific Reports* 15, no. 6 (2016).

Sookias, R. B., Butler, R. J., Benson, R. B. J., "Rise of dinosaurs reveals major body-size transitions are driven by passive processes of trait evolution," *Proceedings of the Royal Society B* 279 (2012): 2180–87.

Wilson, J. A., Mohabey, D. M., Peters, S. E., Head, J. J., "Predation upon hatchling dinosaurs by a new snake from the Late Cretaceous of India," *PLOS Biology* 8, no. 3 (2010).

Chapter 3: Growing Up Dinosaur

Bakker, R., Williams, M., Currie, P., "*Nanotyrannus*, a new genus of pygmy tyrannosaur, from the latest Cretaceous of Montana," *Hunteria* 1, no. 5 (1988): 1–30.

Carr, T. D., "Craniofacial ontogeny in Tyrannosauridae (Dinosauria: Coelurosauria)," *Journal of Vertebrate Paleontology* 19, no. 3 (1999): 497–520.

Dodson, P., "Taxonomic implications of relative growth in lambeosaurine hadrosaurs," *Systematic Zoology* 24, no. 1 (1975): 37–54.

Farke, A. A., Chok, D. J., Herrero, A., Scolieri, B., Werning, S., "Ontogeny in the tube-crested dinosaur Parasaurolophus (Hadrosauridae) and heterochrony in hadrosaurids," *PeerJ* 1 (2013).

Gilmore, C. W., "A new carnivorous dinosaur from the Lance Formation of Montana," *Smithsonian Miscellaneous Collections* 106, no. 13 (1946): 1–19.

Otero, A., Cuff, A. R., Allen, V., Sumner-Ronney, L., Pol, D., and Hutchinson, J. R., "Ontogenetic changes in the body plan of the sauropodomorph dinosaur Mussaurus patagonicus reveal shifts of locomotor stance during growth," *Scientific Reports* 9, no. 1 (2019).

Otero, A., and Pol, D., "Postcranial anatomy and phylogenetic relationships of *Mussaurus patagonicus* (Dinosauria, Sauropodomorpha)," *Journal of Vertebrate Paleontology* 33, no. 5 (2013): 1138–68.

Whitlock, J. A., Wilson, J. A., and Lamanna, M., "Description of a nearly complete juvenile skull of Diplodocus (Sauropoda: Diplodocoidea) from the Late Jurassic of North America," *Journal of Vertebrate Paleontology* 30, no. 2 (2010): 442–57.

Chapter 4: Hot-Running Dinosaurs

Böhme, W., and Nickel, H., "Who was the first to observe parental care in crocodiles?," *Herpetological Bulletin* 74 (2001): 16–18.

Chiarenza, A. A., Mannion, P. D., Farnsworth, A., Carrano, M. T., and Varela, S., "Climatic constraints on the biogeographic history of Mesozoic dinosaurs," *Current Biology* 32, no. 3 (2022): P570–85.

Colinvaux, P., *Why Big Fierce Animals Are Rare: An Ecologist's Perspective* (Princeton University Press, 1979).

Druckenmiller, P., Erickson, G., Brinkman, D., Brown, C., Eberle, J., "Nesting at extreme polar latitudes by non-avian dinosaurs," *Current Biology* 31, no. 16 (2021): P3469–78.E5.

Laskar, A. H., Mohabey, D., Bhattacharya, S. K., Liang, M.-C., "Variable thermoregulation of Late Cretaceous dinosaurs inferred by clumped isotope analysis of fossilized eggshell carbonates," *Heliyon* 6, no. 10 (2020): e05265.

Osborn, H. F., "Ornitholestes hermanni, a new compsognathoid dinosaur from the Upper Jurassic," *Bulletin of the American Museum of Natural History* 19, no. 12 (1903): 459–64.

Thomas, R., Olson, E., eds., *A Cold Look at the Warm-Blooded Dinosaurs* (Westview Press for the American Association for the Advancement of Science, 1980).

Wiemann, J., Menéndez, I., Crawford, J. M., et al., "Fossil biomolecules reveal an avian metabolism in the ancestral dinosaur," *Nature* 606 (2022): 522–26.

Chapter 5: The Largest Creatures to Walk the Earth

Carballido, J., Pol, D., Otero, A., Cerda, I., Salgado, L., Garrido, A., "A new giant titanosaur sheds light on body mass evolution among sauropod dinosaurs," *Proceedings of the Royal Society B* 284, no. 1860 (2017).

D'Emic, M. D., "The evolution of maximum terrestrial body mass in sauropod dinosaurs," *Current Biology* 33, no. 9 (2023): PR349–50.

Hone, D. W. E., and Rauhut, O. W. M., "Feeding behaviour and bone utilization by theropod dinosaurs," *Lethaia* 43, no. 2 (2010): 232–44.

Janis, C. M., Carrano, M., "Scaling of reproductive turnover in archosaurs and mammals: why are large terrestrial mammals so rare?," *Annales Zoologici Fennici* 28, no. 3/4 (1991): 201–16.

Mallon, J., Hone, D., "Estimation of maximum body size in fossil species: a case study using *Tyrannosaurus rex*," *Ecology and Evolution* 14, no. 7 (2024): e11658.

Sander, P. M., Christian, A., Clauss, M., et al., "Biology of the sauropod dinosaurs: the evolution of gigantism," *Biological Reviews* 86, no. 1 (2011): 117–55.

Chapter 6: Dinosaurs of a Feather

Benton, M. J., Currie, P. J., Xu, X., "A thing with feathers," *Current Biology* 31, no. 21 (2021): PR1406–9.

Cincotta, A., Nicolaï, M., Campos, H., McNamara, M., D'Alba, L., Shawkey, M., et al., "Pterosaur melanosomes support signalling functions for early feathers," *Nature* 604, (2022): 684–88.

Godefroit, P., Sinitsa, S., Dhouailly, D., Bolotsky, Y., Sizov, McNamara, M., et al., "A Jurassic ornithischian dinosaur from Siberia with both feathers and scales," *Science* 345, no. 6195 (2014): 451–55.

Li, Q., Gao, K., Vinther, J., Shawkey, M., Clarke, J., D'Alba, L., et al., "Plumage color patterns of an extinct dinosaur," *Science* 327, no. 5971 (2010): 1369–72.

Mayr, G., Peters, S., Plodowski, G., Vogel, O., "Bristle-like integumentary structures at the tail of the horned dinosaur *Psittacosaurus*," *Naturwissenschaften* 89 (2002): 361–65.

Pei, R., Pittman, M., Goloboff, P., Dececchi, T., Habib, M., Kaye, T., et al., "Potential for powered flight neared by most close avialian relatives, but few crossed its thresholds," *Current Biology* 30, no. 20, (2020): 4033–46.e8.

Vinther, J., "Reconstructing vertebrate paleocolor," *Annual Review of Earth and Planetary Sciences* 48 (2020): 345–75.

Vinther, J., Briggs, D. E. G., Prum, R. O., Saranathan, V., "The colour of fossil feathers," *Biology Letters* 4, no. 5 (2008).

Wellnhofer, P., "A short history of research on *Archaeopteryx* and its relationship with dinosaurs," *Geological Society, London, Special Publications* 343 (2010): 237–350.

Zhang, F., Kearns, S. L., Orr, P. J., et al., "Fossilized melanosomes and the colour of Cretaceous dinosaurs and birds," *Nature* 463 (2010): 1075–78.

Chapter 7: Dinosaur Diets

Brown, C. M., Greenwood, D. R., Kalyniuk, J. E., et al., "Dietary palaeoecology of an Early Cretaceous armoured dinosaur (Ornithischia; Nodosauridae) based on floral analysis of stomach contents," *Royal Society Open Science* 7, no. 6 (2020).

Buckland, W., *Geology and Mineralogy Considered with Reference to Natural Theology* (London: William Pickering, 1837).

Chin, K., "The paleobiological implications of herbivorous dinosaur coprolites from the Upper Cretaceous Two Medicine Formation of Montana: why eat wood?" *Palaios* 22, no. 5 (2007): 554–66.

Chin, K., Feldmann, R. M., Tashman, J. N., "Consumption of crustaceans by megaherbivorous dinosaurs: dietary flexibility and dinosaur life history strategies," *Scientific Reports* 7, no. 1 (2017).

Chin, K., Tokaryk, T. T., Erickson, G. M., and Calk, L. C., "A king-size theropod coprolite," *Nature* 393 (1998): 680–82.

D'Emic, M. D., Whitlock, J. A., Smith, K. M., Fisher, D. C., and Wilson, J. A., "Evolution of high tooth replacement rates in sauropod dinosaurs," *PLOS ONE* 8, no. 7 (2013).

Hone, D. W. E., and Chure, D. J., "Difficulties in assigning trace makers from theropodan bite marks: an example from a young diplodocoid sauropod," *Lethaia* 51, no. 3 (2018): 456–66.

Hone, D. W. E., and Watabe, M., "New information on scavenging and selective feeding behaviour of tyrannosaurids," *Acta Palaeontologica Polonica* 55, no. 4 (2010): 627–34.

Molnar, R. E., and Clifford, H. T., "Gut contents of a small ankylosaur," *Journal of Vertebrate Paleontology* 20, no. 1 (2000): 194–96.

Therrien, F., Zelenitsky, D. K., Tanaka, K., et al., "Exceptionally preserved stomach contents of a young tyrannosaurid reveal an ontogenetic dietary shift in an iconic extinct predator," *Science Advances* 9, no. 49 (2023).

Torices, A., Wilkinson, R., Arbour, V. M., Ruiz-Omeñaca, J. I., and Currie, P. J., "Puncture-and-pull biomechanics in the teeth of predatory coelurosaurian dinosaurs," *Current Biology* 28, no. 9 (2018): P1467–74.

Wings, O., "The rarity of gastroliths in sauropod dinosaurs—a case study in the Late Jurassic Morrison Formation, western USA," *Fossil Record* 18, (2014): 1–16.

Wings, O., and Sander, P. M., "No gastric mill in sauropod dinosaurs: new evidence from analysis of gastrolith mass and function in ostriches," *Proceedings of the Royal Society B* 274, no. 1610 (2007).

Zheng, X., Wang, X., Sullivan, C., et al., "Exceptional dinosaur fossils reveal early origin of avian-style digestion," *Scientific Reports* 8, no. 1 (2018): 14217.

Chapter 8: How Dinosaurs Made Their World

Bakker, R. T., "Dinosaur physiology and the origin of mammals," *Evolution* 25, no. 4 (1971): 636–58.

Brocklehurst, N., Panciroli, E., Benevento, G. L., Benson, R. B. J., "Mammaliaform extinctions as a driver of the morphological radiation of Cenozoic mammals," *Current Biology* 31 no. 13 (2021): P2955–63.

Carvalho, I. S., Leonardi, G., Rios-Netto, A. M., et al., "Dinosaur trampling from the Aptian of Araripe Basin, NE Brazil, as tools for paleoenvironmental interpretation," *Cretaceous Research* 117, no. 104626 (2021).

Carvalho, M., Jaramillo, C., La Parra, F., Caballero-Rodríguez, D., Herrera, F., Wing, S., et al., "Extinction at the end-Cretaceous and the origin of modern Neotropical rainforests," *Science* 372, no. 6537, (2021): 63–68.

Chin, K., and Gill, B. D., "Dinosaurs, dung beetles, and conifers: participants in a Cretaceous food web," *Palaios* 11, no. 3 (1996): 280–85.

Han, G., Mallon, J. C., Lussier, A. J., Wu, X.-C., Mitchell, R., and Li, L.-J., "An extraordinary fossil captures the struggle for existence during the Mesozoic," *Scientific Reports* 13 (2023): 11221.

"History of Paleomammology: The first Mesozoic Mammals—or—the Platypus and the Phascolotherium," History of Geology, August 4, 2010, historyofgeology.fieldofscience.com/2010/08/history-of-paleomammology-first.html.

Hu, H., Wang, Y., McDonald, P. G., et al., "Earliest evidence for fruit consumption and potential seed dispersal by birds," *eLife* 11 (2022).

Hu, Y., Meng, J., Wang, Y., Li, C., "Large Mesozoic mammals fed on young dinosaurs," *Nature* 433 (2005): 149–52.

Lull, R. S., *Organic Evolution* (Macmillan, 1921).

Luo, Z., "Transformation and diversification in early mammal evolution," *Nature* 450, no. 7172 (2007): 1011–19.

Mallon, J. C., and Anderson, J. S., "Skull ecomorphology of megaherbivorous dinosaurs from the Dinosaur Park Formation (Upper Campanian) of Alberta, Canada," *PLOS ONE* 8, no. 7 (2013).

Mallon, J. C., Evans, D. C., Ryan, M. J., and Anderson, J. S., "Feeding height stratification among the herbivorous dinosaurs from the Dinosaur Park Formation (Upper Campanian) of Alberta, Canada," *BMC Ecology* 13, no. 14 (2013).

O'Connor, J., Clark, A., Herrera, F., et al., "Direct evidence of frugivory in the Mesozoic bird Longipteryx contradicts morphological proxies for diet," *Current Biology* 34, no. 19 (2024): P4559–66.

Romano, M., and Manucci, F., "Resizing Lisowicia bojani: volumetric body mass estimate and 3D reconstruction of the giant Late Triassic dicynodont," *Historical Biology* 33, no. 4 (2021): 474–79.

Simpson, E. L., Hilbert-Wolf, H. L., Wizevich, M. C., et al., "Predatory digging behavior by dinosaurs," *Geology* 38, no. 8 (2010): 699–702.

Sloan, R. E., Rigby Jr., J. K., Van Valen, L. M., and Gabriel, D., "Gradual dinosaur extinction and simultaneous ungulate radiation in the Hell Creek Formation," *Science* 232, no. 4750 (1986): 629–33.

Smith, F., Lyons, S., "How big should a mammal be? A macroecological look at mammalian body size over space and time," *Philosophical Transactions of the Royal Society B* 366, no. 1576 (2011).

Thulborn, T., "Impact of sauropod dinosaurs on lagoonal substrates in the Broome Sandstone (Lower Cretaceous), Western Australia," *PLOS ONE* 7, no. 5 (2012).

Wieland, G. R., "Dinosaur extinction," *The American Naturalist* 59, no. 665 (1925).

Wu, Y., Ge, Y., Han, H., et al., "Intra-gastric phytoliths provide evidence for folivory in basal avialans of the Early Cretaceous Jehol Biota," *Nature Communications* 14, no. 1 (2023).

Chapter 9: Decoration and Defense

Arbour, V. M., Zanno, L. E., and Evans, D. C., "Palaeopathological evidence for intraspecific combat in ankylosaurid dinosaurs," *Biology Letters* 18, no. 12 (2022).

Farke, A., "Evaluating combat in ornithischian dinosaurs," *Journal of Zoology* 292, no. 4 (2014): 242–49.

Farke, A., "Horn use in Triceratops (Dinosauria: Ceratopsidae): testing behavioral hypotheses using scale models," *Palaeontologia Electronica* 7, no. 1 (2004): 1–10.

Farlow, J., Hayashi, S., Tattersall, G., "Internal vascularity of the dermal plates of *Stegosaurus* (Ornithschia, Thyreophora)," *Swiss Journal of Geosciences* 103, (2010): 173–85.

Gould, S., "Darwinism and the expansion of evolutionary theory," *Science* 216, no. 4544 (1982): 380–87.

Hone, D., Naish, D., Cuthill, I., "Does mutual sexual selection explain the evolution of head crests in pterosaurs and dinosaurs?," *Lethaia* 45, no. 2, (2011): 139–56.

Nabavizadeh, A., "How *Triceratops* got its face: an update on the functional evolution of the ceratopsian head," *The Anatomical Record* 306, no. 7 (2023): 1951–68.

Peterson, J., Henderson, M., Scherer, R., Vittore, C., "Face biting on a juvenile tyrannosaurid and behavioral implications," *PALAIOS* 24, no. 11: 780–84.

Tanke, D., Currie, P., "Head-biting behavior in theropod dinosaurs: paleopathological evidence," *GAIA* (1998): 167–84.

Woofruff, D., Ackermans, N., "Headbutting through time: a review of this hypothesized behavior in 'dome-headed' fossil taxa," *The Anatomical Record* (2024).

Chapter 10: The Social Dinosaur

Chiba, K., Ryan, M. J., Braman, D. R., et al., "Taphonomy of a monodominant *Centrosaurus apertus* (Dinosauria: Ceratopsia) bone bed from the upper Oldman Formation of southeastern Alberta," *PALAIOS* 30, no. 9 (2015): 655–67.

Kim, K. S., Lockley, M. G., Lim, J. D., Buckley, L., and Xing, L., "Small scale scrapes suggest avian display behavior by diminutive Cretaceous theropods," *Cretaceous Research* 66 (2016): 1–5.

Kobayashi, Y., and Lü, J.-C., "A new ornithomimid dinosaur with gregarious habits from the Late Cretaceous of China," *Acta Palaeontologica Polonica* 48, no. 2 (2003): 235–59.

Lockley, M. G., McCrea, R. T., Buckley, L. G., et al., "Theropod courtship: large scale physical evidence of display arenas and avian-like scrape ceremony behaviour by Cretaceous dinosaurs," *Scientific Reports* 6, no. 1 (2016).

Mallon, J., Holmes, R., Eberth, D., Ryan, M., Anderson, J., "Variation in the skull of *Anchiceratops* (Dinosauria, Ceratopsidae) from the Horseshoe Canyon Formation (Upper Cretaceous) of Alberta," *Journal of Vertebrate Paleontology* 31, no. 5 (2011): 1047–71.

Mathews, J. C., Brusatte, S. L., and Williams, S. A., "The first Triceratops bone bed and its implications for gregarious behavior," *Journal of Vertebrate Paleontology* 29, no. 1 (2009): 286–90.

McCrea, R. T., Buckley, L. G., Farlow, J. O., et al., "A 'terror of tyrannosaurs': the first trackways of tyrannosaurids and evidence of gregariousness and pathology in tyrannosauridae," *PLOS ONE* 10, no. 1 (2014).

Myers, T. S., and Fiorillo, A. R., "Evidence for gregarious behavior and age segregation in sauropod dinosaurs," *Palaeogeography, Palaeoclimatology, Palaeoecology* 274, nos. 1–2 (2009): 96–104.

Osborn, H., "Tyrannosaurus: restoration and model of the skeleton," *Bulletin of the AMNH* 32, no. 4 (1913): 91–92.

Peterson, J. E., Tseng, Z. J., and Brink, S., "Bite force estimates in juvenile Tyrannosaurus rex based on simulated puncture marks," *PeerJ* 9 (2021).

Roach, B. T., and Brinkman, D. L., "A reevaluation of cooperative pack hunting and gregariousness in Deinonychus antirrhopus and other nonavian theropod dinosaurs," *Bulletin of the Peabody Museum of Natural History* 48, no. 1 (2007): 103–38.

Ryan, M. J., Russell, A. P., Eberth, D. A., and Currie, P. J., "The taphonomy of a Centrosaurus (Ornithischia: Certopsidae) bone bed from the Dinosaur Park Formation (Upper Campanian), Alberta, Canada, with comments on cranial ontogeny," *PALAIOS* 16, no. 5 (2001): 482–506.

Chapter 11: Dinosaurs Undone

Allentoft, M., Collins, M., Harker, D., Haile, J., Oskam, C., Hale, M., et al., "The half-life of DNA in bone: measuring decay kinetics in 158 dated fossils," *Proceedings of the Royal Society B* 279, no. 1748 (2012).

Alvarez, L., Alvarez, W., Asaro, F., Michel, H., "Extraterrestrial cause of the Cretaceous-Tertiary extinction," *Science* 208, no. 4448 (1980): 1095–108.

Benton, M., "Scientific methodologies in collision: the history of the study of the extinction of the dinosaurs," *Evolutionary Biology* 24 (1990): 371–400.

Chiarenza, A. A., Farnsworth, A., and Mannion, P. D., "Asteroid impact, not volcanism, caused the end-Cretaceous dinosaur extinction," *PNAS* 117, no. 29 (2020): 17084–93.

Gulick, S., Bralower, T., Ormö, J., Hall, B., Grice, K., Schaefer, B., et al., "The first day of the Cenozoic," *PNAS* 116, no. 39 (2019): 19342–51.

Larson, D. W., Brown, C. M., and Evans, D. C., "Dental disparity and ecological stability in bird-like dinosaurs prior to the end-Cretaceous mass extinction," *Current Biology* 26, no. 10 (2016): P1325–33.

Robertson, D., McKenna, M., Toon, O., Hope, S., Lillegraven, J., "Survival in the first hours of the Cenozoic," *GSA Bulletin* 116 (2004): 760–68.

Schulte, P., Alegret, L., Arenillas, I., et al., "The Chicxulub asteroid impact and mass extinction at the Cretaceous-Paleogene boundary," *Science* 327, no. 5970 (2010): 1214–18.

Schultz, T. R., Sosa-Calvo, J., Kweskin, M. P., et al., "The coevolution of fungus-ant agriculture," *Science* 386, no. 6717 (2024): 105–10.

Chapter 12: How to Become a Fossil

Darwin, C., *On the Origin of Species by Means of Natural Selection*, Sixth Edition (London: John Murray, 1872).

Delclòs, X., Peñalver, E., Barrón, E., et al., "Amber and the Cretaceous resinous interval," *Earth-Science Reviews* 243 (2023).

Drumheller, S. K., Boyd, C. A., Barnes, B. M. S., and Householder, M. L., "Biostratinomic alterations of an Edmontosaurus 'mummy' reveal a pathway for soft tissue preservation without invoking 'exceptional conditions,'" *PLOS ONE* 17, no. 10 (2022).

Leach, C., Hoffman, E., Dodson, P., "The promise of taphonomy as a nomothetic discipline: taphonomic bias in two dinosaur-bearing faunas in North America," *Canadian Journal of Earth Sciences* 58, no. 9 (2021).

Martinez-Delclòs, X., Briggs, D. E. G., and Peñalver, E., "Taphonomy of insects in carbonates and amber," *Palaeogeography, Palaeoclimatology, Palaeoecology* 203, nos. 1–2 (2004): 19–64. Osborn, H. F., "The epidermis of an iguanodont dinosaur," *Science* 29, no. 750 (1909): 793–95.

Riga, B., Casal, G., Fiorillo, A., David, L., "Taphonomy: overview and new perspectives related to the paleobiology of giants," in, Otero, A., Carballido, J., Pol, D., eds., *South American Sauropodmorph Dinosaurs* (Springer, 2022).

Schweitzer, M., Zheng, W., Dickinson, E., Scannella, J., Hartstone-Rose, A., Sjövall, P., et al., "Taphonomic variation in vascular remains from Mesozoic non-avian dinosaurs," *Scientific Reports* 15, no. 4359 (2025).

Wang, S., Dodson, P., "Estimating the diversity of dinosaurs," *PNAS* 103, no. 37 (2006): 13601–5.

Xu, X., and Norell, M. A., "A new troodontid dinosaur from China with avian-like sleeping posture," *Nature* 431 (2004): 838–41.

Afterword

Lee, Y., Barsbold, R., Currie, P., et al., "Resolving long-standing enigmas of a giant ornithomimosaur *Deinocheirus mirificus*," *Nature* 515: 257–60.

Image Credits

Page x: Public domain.

Page 7: Public domain.

Page 10: Public domain.

Page 13: Courtesy of the Museum für Naturkunde Berlin, CC BY 4.0.

Page 16: Illustration by Bálint Benke/paleostock.com.

Page 17: Diagram by Scott Hartman, CC BY 2.5.

Page 26: From Jeffrey W. Martz and Bryan J. Small, "Non-dinosaurian dinosauromorphs from the Chinle Formation (Upper Triassic) of the Eagle Basin, northern Colorado: Drommomeron romeri (Lagerpetidae) and a new taxon, Kwanasaurus williamparkeri (Silesauridae)," *PeerJ* 7 (2019): e7551, DOI: 10.7717/peerj.7551/fig-24, CC BY 4.0.

Page 28: Created by Beth Bugler, based on diagram by Dr. Thomas Holtz.

Page 29: Art by Brian Engh, CC BY 2.5.

Page 30: Diagrams by AdmiralHood, CC BY-SA 3.0.

Page 36: Art by Heather Kyoht Luterman, CC BY 2.5.

Page 41: Wiemann et al., "The blue-green eggs of dinosaurs: How fossil metabolites provide insights into the evolution of bird reproduction," *PeerJ Preprints* 3 (2015): e1323, DOI: 10.7287/peerj.preprints.1080v1, CC BY-SA 4.0.

Page 45: Albert Prieto-Marquez and Merrile Guenther, "Perinatal specimens of Maiasaura from the Upper Cretaceous of Montana (USA): Insights into the early ontogeny of saurolophine hadrosaurid dinosaurs," *PeerJ* 6 (2018): e4734, DOI: 10.7717/peerj.4734, CC BY 4.0.

Page 48: E. Martín Hechenleitner et al., "What do giant titanosaur dinosaurs and modern Australasian megapodes have in common?," *PeerJ* 3 (2015): e1341, DOI: 10.7717/peerj.1341, CC BY 4.0.

Page 50: Art by FunkMonk (Michael B. H.), CC BY-SA 3.0.

Page 56: Andrew A. Farke et al., "Ontogeny in the tube-crested dinosaur *Parasaurolophus* (Hadrosauridae) and heterochrony in hadrosaurids," *PeerJ* 1 (2013): e182, DOI: 10.7717/peerj.182/fig-3, CC BY 4.0.

Page 59: Art by A. Atuchin, from D. C. Woodruff et al., "The Smallest Diplodocid Skull Reveals Cranial Ontogeny and Growth-Related Dietary Changes in the Largest Dinosaurs," *Scientific Reports* 8 (2018): 14341, DOI: 10.1038/s41598-018-32620-x, 2018, CC BY 4.0.

Page 61: Ignacio Alejandro Cerda et al., "Novel insight into the origin of the growth dynamics of sauropod dinosaurs," *PLoS ONE* 12, no. 6 (2017): e0179707, DOI: 10.1371/journal.pone.0179707.g010, CC BY 4.0.

Page 64: Art by Zissoudisctrucker, CC BY-SA 4.0.

Page 65: C. M. Bullar et al., "Ontogenetic braincase development in *Psittacosaurus lujiatunensis* (Dinosauria: Ceratopsia) using micro-computed tomography," *PeerJ* 7 (2019): e7217, DOI: 10.7717/peerj.7217/fig-1, CC BY 4.0.

Page 72: Public domain.

Page 74: Art by PaleoNeolitic, CC BY-SA 4.0.

Page 77: Fred Wierum, CC BY-SA 4.0.

Page 80: Illustration by Rudolf Hima/paleostock.com.

Page 82: Andrey Atuchin, "The first juvenile dromaeosaurid (Dinosauria: Theropoda) from Arctic Alaska," *PLoS ONE* 15, no. 7 (2020): e0235078, DOI: 10.1371/journal.pone.0235078, CC BY 2.5.

Page 86: Courtesy Rob Glover from Bradford, UK, CC BY-SA 2.0.

Page 88: Art by Fred Wierum, CC BY-SA 4.0.

Page 90: M. P. Taylor and M. J. Wedel, "Why sauropods had long necks; and why giraffes have short necks," *PeerJ* 1 (2013): e36, DOI: 10.7717/peerj.36/fig-3, CC BY 4.0.

Page 95: Illustration by Sergey Krasovskiy/paleostock.com.

Page 98: O. W. M. Rauhut et al., "The oldest *Archaeopteryx* (Theropoda: Avialiae): a new specimen from the Kimmeridgian/Tithonian boundary of Schamhaupten, Bavaria," *PeerJ* 6 (2018): e4191, DOI: 10.7717/peerj.4191/fig-4, CC BY 4.0.

Page 101: Illustration by Julio Lacerda/paleostock.com.

Page 102: J. Benito et al., "Forty new specimens of *Ichthyornis* provide unprecedented insight into the postcranial morphology of crownward stem group birds," *PeerJ* 10 (2022): e13919, DOI: 10.7717/peerj.13919/fig-33, CC BY 4.0.

Page 104: S. Hartman et al., "A new paravian dinosaur from the Late Jurassic of North America supports a late acquisition of avian flight," *PeerJ* 7 (2019): e7247, DOI: 10.7717/peerj.7247/fig-5, CC BY 4.0.

Page 107: D. W. E. Hone et al., "The Extent of the Preserved Feathers on the Four-Winged Dinosaur *Microraptor gui* under Ultraviolet Light," *PLoS ONE* 5, no. 2 (2010): e9223, DOI: 10.1371/journal.pone.0009223, CC BY 2.5.

Page 110: Q. Li et al., "Elaborate plumage patterning in a Cretaceous bird," *PeerJ* 6 (2018): e5831, DOI: 10.7717/peerj.5831/fig-6, CC BY 4.0.

Page 114: C. M. Brown et al., "Rare evidence for 'gnawing-like' behavior in a small-bodied theropod dinosaur," *PeerJ* 9 (2021): e11557, DOI: 10.7717/peerj.11557/fig-5, CC BY 4.0.

Page 116: L. Xing et al., "Abdominal Contents from Two Large Early Cretaceous Compsognathids (Dinosauria: Theropoda) Demonstrate Feeding on Confuciusornithids and Dromaeosaurids," *PLoS ONE* 7, no. 8 (2012): e44012, DOI: 10.1371/journal.pone.0044012, CC BY 2.5.

Page 118: Carol Abraczinskas et al., "Structural Extremes in a Cretaceous Dinosaur," *PLoS ONE* 2, no. 11 (2007): e1230, DOI: 10.1371/journal.pone.0001230, CC BY 2.5.

Page 126: Illustration by Andrey Atuchin/paleostock.com.

Page 128: T. Thulborn, "Impact of Sauropod Dinosaurs on Lagoonal Substrates in the Broome Sandstone (Lower Cretaceous), Western Australia," *PLoS ONE* 7, no. 5 (2012): e36208, DOI: 10.1371/journal.pone.0036208.g010, CC BY 4.0.

Page 130: Image courtesy of J. T. Csotonyi, CC BY 2.5.

Page 133: Art by Michael Skrepnick, from Gang Han et al., "An extraordinary fossil captures the struggle for existence during the Mesozoic," *Scientific Reports* 13 (2023): 11221 (fig. 2), DOI: 10.1038/s41598-023-37545-8, CC BY 4.0.

Page 135: Art by Kaek, CC BY 3.0.

Page 137: Public domain.

Page 140: Andrew Farke and Ewan Wolff, "Evidence of Combat in Triceratops," *PLoS ONE* 4 (2009): e4252, DOI: 10.1371/journal.pone.0004252.g002, CC BY 4.0.

Page 142: D. J. Chure and M. A. Loewen, "Cranial anatomy of *Allosaurus jimmadseni*, a new species from the lower part of the Morrison Formation (Upper Jurassic) of Western North America," *PeerJ* 8 (2020): e7803, DOI: 10.7717/peerj.7803/fig-3, CC BY 4.0.

Page 144: Illustration by Sydney Mohr, from V. M. Arbour et al., "A New Ankylosaurid Dinosaur from the Upper Cretaceous (Kirtlandian) of New Mexico with Implications for Ankylosaurid Diversity in the Upper Cretaceous of Western North America," *PLoS ONE* 9, no. 9 (2014): e108804, DOI: 10.1371/journal.pone.0108804, CC BY 2.5.

Page 146: Illustration by Ryan Steiskal, CC BY 2.5.

Page 147: Modified after Scott D. Sampson et al., "New Horned Dinosaurs from Utah Provide Evidence for Intracontinental Dinosaur Endemism," *PLoS ONE* 5 (2010): e12292, DOI: 10.1371/journal.pone.0012292.g008, CC BY 2.5.

Page 151: Public domain.

Page 154: Martin G. Lockley et al., "Theropod courtship: large scale physical evidence of display arenas and avian-like scrape ceremony behaviour by Cretaceous dinosaurs," *Nature* (2016), DOI: 10.1038/srep18952, CC BY 4.0.

Page 158: Richard T. McCrea et al., "A 'Terror of Tyrannosaurs': The First Trackways of Tyrannosaurids and Evidence of Gregariousness and Pathology in Tyrannosauridae," *PLoS ONE* 9, no. 7 (2014): e103613, DOI: 10.1371/journal.pone.0103613, CC0 1.0.

Page 159: Art by Sergey Krasovskiy, CC BY 4.0.

Page 160: Art by ABelov2014, CC BY 3.0.

Page 165: Terry A. Gates et al., "Mountain Building Triggered Late Cretaceous North American Megaherbivore Dinosaur Radiation," *PLoS ONE* 7, no. 10 (2012), DOI: 10.1371/journal.pone.0042135, CC BY 4.0.

Page 170: Nicholas R. Longrich and Daniel J. Field, "*Torosaurus* Is Not *Triceratops*: Ontogeny in Chasmosaurine Ceratopsids as a Case Study in Dinosaur Taxonomy," *PLoS ONE* 7, no. 2 (2012): e32623, DOI: 10.1371/journal. pone.0032623, CC BY 4.0.

Page 175: Illustration by Rudolf Hima/paleostock.com.

Page 178: Image by Patrick Lynch, Yale University.

Page 183: Kristian Remes et al., "A New Basal Sauropod Dinosaur from the Middle Jurassic of Niger and the Early Evolution of Sauropoda," *PLoS ONE* 4, no. 9 (2009): e6924, DOI: 10.1371/journal.pone.0006924, CC BY 4.0.

Page 186: Andrew A. Farke et al., "A Ceratopsian Dinosaur from the Lower Cretaceous of Western North America, and the Biogeography of Neoceratopsia," *PLoS ONE* 9, no. 12 (2014): e112055, DOI: 10.1371/journal.pone.0112055, CC BY 4.0.

Page 189: Stephanie K. Drumheller et al., "Biostratinomic alterations of an *Edmontosaurus* 'mummy' reveal a pathway for soft tissue preservation without invoking 'exceptional conditions,'" *PLoS ONE* 17, no. 10 (2022): e0275240, DOI: 10.1371/journal.pone.0275240, CC BY 4.0.

Page 192: Caleb M. Brown, "An exceptionally preserved armored dinosaur reveals the morphology and allometry of osteoderms and their horny epidermal coverings," *PeerJ* 5 (2017): e4066, DOI: 10.7717/peerj.4066, CC BY 4.0.

Page 197: Illustration by Corbin Rainbolt, Studio 252MYA/paleostock.com.

Page 200: Art by PaleoNeolitic, CC BY 4.0.

Acknowledgments

The ongoing history of Dinosauria is a massive subject. The story not only involves the "terrible lizards" and their relationship with the rest of our planet since the Triassic, but how our ideas about them have continued to change. Even now, as you're reading this, paleontologists are searching for, discovering, and describing new fossils that will further change how we understand *Tyrannosaurus* and family. Experts will find new species, reassess old finds, and otherwise adjust what we know today as the science unfolds. The effort is endlessly fascinating, but also poses a challenge in writing a crash course on animals that we are still just beginning to truly understand.

I'm grateful to editor Nick Cizek at The Experiment for throwing me the challenge in the first place. Even though I've been writing books about dinosaurs and other prehistoric creatures for more than fifteen years, I'd never had the chance to write my own version of a Dinosaurs 101 course. Nick guided this manuscript through several iterations and helped keep the balance between the history of the dinosaurs themselves and the people who have studied them over the past two centuries.

My agent, Deirdre Mullane, has been a tireless advocate of my work and helped me maintain my balance as I've kept several plates spinning at once. Her diligent work helped bring this book about at just the right time. I'm also thankful to the many paleontologists I have ventured into the field with, interviewed, chatted with at conferences, and have otherwise informed the chapters in this book. There are too many of you to mention by name, but I hope that you see your work reflected in the ideas and references contained within this book.

As much as I love my dinosaurs, though, I do need a break from them sometimes. This book wouldn't exist without the care and support of my friends, who have cheered me on through the process of writing this history. Oliver, June, Alex, Becca, Blue, London, Alphonse, Ezekiel, and Zandra, you have all endured my irrepressible dinomania and have reminded me to take care of myself as well as the text on these pages.

Most of all, I must thank my girlfriend Splash and our furry family—Jet the German shepherd and Hobbes, Strata, and Joey, our cats. Splash has loved and supported my work before we even met, and dinosaurs are one of our shared points of affection. From fossil road trips to chatting about the latest finds, we've fostered our love of the prehistoric with each other. And while the pets have no idea why I sit for hours on end at my desk making little clicking sounds, I've been glad for all the time they've spent curled up nearby and have reminded me to take play breaks with them. I can only write books like this from the warmth of home, and all of them keep it a very loving place.

Thank you so much for reading this book. I hope you've found wonder as well as encouragement to dig in further for yourself. Whether it's in museums, public fossil sites, or other books, I hope you keep chasing dinosaurs and all we're learning about these magnificent Mesozoic creatures.

Index

NOTE: Page references in *italics* refer to figures, illustrations, and photos.

Acheroraptor, *175*, 176
Agilodocodon, 135
Akainacephalus, 145
Alamosaurus, 95, 159–60, *159*
Alaska, 82–83, *82*
Alberta, 55–56, 57, 62, 130–31, 138, 161, 162, 174, 182
alligators, 33, 72–73, 143, 153
Allosaurus, 31, 44, 47, 78, 86, 94, 117–18, *142*, 193–94
American Museum of Natural History, 12, 46, 151–52, *151*
Amphicoelias fragillimus, 95
Anatosaurus, 55
Anatotitan, 55
Anchiornis, 105, 110
Ankylosaurus, 31, 82, 145
Anning, Mary, 109
Antarctica, xi
Apatosaurus, 29, *29*, 49–50, 70, 76, 86–87, 91, 94, 116–18
Appalachia, 161–62
Aquilops, *186*
Araripe Basin, 127–28
Arbour, Victoria, 145–47
Archaeopteryx, 14, *98*, 100–103, *101*, *104*, 105–6, 108
Arctic Circle, *82*, 83
Argentina, x, 25, 47, 58
Argentinosaurus, 29, 87, 95
Arizona, 25, 143
armor, 143–49, *144*, *146*, 199
asteroid event, 168–74, 176, 179
Australia, xi, 123, 128, *128*, 188

"Baby Louie," 46–47
Barosaurus, 86, 87

Bavaria, 99–100
Belgium, 74
Benton, Michael, 167
Big Bend National Park, 159–60
birds: anatomical traits, 30–31, *30*, 91–92; diet of, 177; earliest birds, 103, *110*; evolution of, 100–102, *101–2*, 104, *104*, 177; family tree, 1–2, 31, 100–103, *101–2*, 104, *104*, 105; habitats, 129–30; as living dinosaurs, 103, 165, *175*, 176–78, 179; reproductive behavior, 42–43
Bison alticornis, 138
Bone Cabin Quarry, 73–74
Borealopelta, 123–24, *192*
Brachiosaurus, 25, 70, 75, 81, 87, 92–93
Brachyceratops, 58
Brandvold, Marion, 50
Brazil, 2, 127–28
breeding behavior. *See* eggs and nests
B-rex, 43–44
Bridgewater Treatises, 113
British Columbia, 152, *158*
Brown, Barnum, 151
Bruhathkayosaurus, 95
Buckland, Mary, 109
Buckland, William, 7–8, *7*, 113–14, 184

Caihong, 111
Camarasaurus, 78, 86, *86*, 87, 94, 119
Carnegie Museum of Natural History, 12–13, 87, 152
Carnegie Quarry of Dinosaur National Monument, 186
Carvalho, Ismar de Souza, 128
Central Atlantic Magmatic Province (CAMP), 35
Centrosaurus, 55, 139–40, 141, 155, *160*, 161, 162

Ceratosaurus, 117–18, 194
Charig, Alan, 26, 27
Chasmosaurus, 140–41
Chen, Peiji, 103
Chiarenza, Alfio Alessandro, 81
Chin, Karen, 115, 122, 123, 126–27
China, xi, 22, *80*, 97–98, 103–4, 129–30, 186
Christman, Erwin, 151–52
Claosaurus, 55
Cleveland-Lloyd Dinosaur Quarry, 47
Coelophysis, 25, 33
A Cold Look at Hot-Blooded Dinosaurs (conference and book), 79–80
colonialism, 4, 10–13, *13*
Colonna, Fabio, 3
Colorado, 12, 47, 49, 85–86, 117, 137–39, *137*, 154, 174
Confuciusornis, *110*
conifer trees, 34, 120, 122–24, 126–27, *126*, 191
Cope, Edward Drinker, 11–12
Copper Ridge, 193
coprolites, 114–15, *114*, 117, 121–24, 126–27
Corral Bluffs, 174
Corythosaurus, 55
courtship displays, 17, 108, 153–55, *154*
crocodiles and relatives: anatomical traits, 33; armor of, 143–44; comparison to dinosaurs, 33, 34–35; diets of, 49, 119; fossils mislabeled as, 4–5, 7, 8, 14; hunting behavior, 157; during Jurassic period, 35–36; misconceptions about, 69; reproductive behavior, 44, 72–73; in "ruling reptiles" group, 24; social behavior, 153; during Triassic period, 26–27, 33, 34–36
Cryolophosaurus, xi
Currie, Phil, 63
Cuvier, Georges, 4, 7, 8

"Dakota," 189–91, *189*
Darwin, Charles, 13–14, 99–101, 105, 166, 182
Daspletosaurus, 65
defense mechanisms: armor as, 143–49, *144*, *146*, 199; future research directions, 199; horns as, 137–40, *137*, *140*, *142*, 146–47,

148; juvenile social groups, 159–61, *159–60*; misconceptions about, 138–39, 140–41, *142*, 146–49, *147*; size as, 94
Deinocheirus, 31, 200–1, *200*
Deinonychus, 15, 31, 76, *77*, 102, 105, 155–58
diet and digestive system of dinosaurs, 113–24; anatomical considerations, 59–62, *59*, 113–21, *114*, *118*, *126*; comparison to mammals, 92–93; diversity of diets, 121–23, 126–27, *126*, 128–31, *130*; and ecosystems, 126–27, *126*, 128–31; feces (coprolite), 114–15, *114*, 117, 121–24, 126–27; and mass extinction survival, 177; stomach contents (fossilized), 115–16, *116*, 117, 124; thermoregulation evidence through, 77–78
Dilophosaurus, *36*, 143
Dimetrodon, 20
dinosaur family tree (Dinosauria): "dinosaur" defined by, 30–32; first scientific name, x, 7; during Jurassic period, 36–37; lineage, 1–2, *28*, 31, 100–3, *101–2*, 104, *104*, 105 (See also birds); misidentifications, 53–59, *64–65* (See also ontogeny); new species identification process, 53–55; origin of Dinosauria name, 8; during Triassic period, 36
Dinosaur National Monument, 85–86, *86*, 117–18
Dinosaur Park Formation, *130*
Dinosaur Provincial Park, 130–31
Dinosaur Renaissance, 15–16, *77*, 78, 103, 155
dinosaurs: age determination, 44, 45–46, *61*; anatomical traits, 15, 30–32, *30*, 91, 102, 116, 117–18, 120; ancestors, 1–2, 19–25, 27, 28–29, *28* (See also dinosaur family tree); colors of, 17, 108–11, *192*, 198; decoration and defense, 137–49 (See also defense mechanisms; ornamentation); defining "what is a dinosaur," 29–33; diets of, 113–24 (See also diet and digestive system of dinosaurs); evidence of, 181–95 (See also fossilization; paleontology); fate of, 165–79 (See also "extinction"

of dinosaurs); as feathered creatures, 97–111 (*See also* feathers and feathery dinosaurs); fossils (overview), x–xi; growth changes, 53–67 (*See also* ontogeny); hot-running dinosaurs, 69–84 (*See also* physiology); largest dinosaurs, 85–96 (*See also* large size of dinosaurs); oldest dinosaur, 24, 25–29; present-day and future research directions, 16–18, *16–17*, 84, 197–202; reproductive strategies, 39–51 (*See also* eggs and nests); rise of dinosaurs, 19, 34–37; social interactions, 151–63 (*See also* social behavior); timeline, viii–ix; world changing by, 125–36 (*See also* ecosystems impacted by dinosaurs)

Diplodocus, 12, 31, *59*, 60–62, 86–87, *88*, *90*, 94–95, 117–19

Diplodocus hallorum, 95

Dodson, Peter, 56

Douglass, Earl, 87

Dracorex, 58

Drumheller, Stephanie, 190–91

Dryosaurus, 49

Dryptosaurus, 71

ecosystems impacted by dinosaurs, 125–36; co-existence of giant plant-eaters, 130–31, *130*; foraging behavior, 126–27; future research directions, 198; misconceptions about, 125–26, 131–36, *133*, *135*; walking, 127–28, *128*

Edmontosaurus, 31, 55, 120, 121, 126, 189–91, *189*, 201

eggs and nests, 39–51; calcium sources for eggs, 43, 123, 143; climate estimates based on, 81; diversity of, 51; fossils of, 39–42, *41*, 46–47; hatching, 51; incubation period, 49; nesting habitats and site fidelity, 47–48, *48*, 89; parental behavior, *41*, 46–47, 49–50, *50*; reproductive advantage of, 42, 44, 46, 89; reproductive age, 42–44; thermoregulation evidence through, 78; vulnerability of nests and young, 45–46, *45*, 49

end-Cretaceous extinction, 165, *165*, 168–79, *175*, *178*

end-Permian mass extinction, 19–20, 21–23

England, x, 6–9, 11

Eoraptor, 25, 26–27

evolution, theory of, 13–16, 166

"extinction" of dinosaurs, 165–79; acceptance of, 4; birds as exception to, 103, 165, *175*, 176–78, 179; end-Cretaceous extinction evidence, *165*, 168–77; misconceptions about, 165–68

Falconer, Hugh, 99–100

Farke, Andy, 57–58

feathers and feathery dinosaurs, 97–111; colors of, 108–11; diets of, 128–29; evolution of feathers, 31, 35, 102–4, 106–7; "extinction" of, 176–77; physiological functions of, *80*, 107–8; for social communication, 108; transitional fossils with feathers, 97–100; wing feathers and flight, 105, *107*, 108. *See also* birds

feces (coprolite), 114–15, *114*, 117, 121–24, 126–27

Field Museum, 12

footprints, *128*, *158*, 193–95

fossilization, 181–95; in amber, 191–92; decomposition prior to, 189–92, *189*, *192*; imperfections of fossil record, 199–200; potential for, 183–87, *186*; process of, 187–88; rarity of, 181–83, *183*, *186*, 189, *197*; trace fossils, *128*, *158*, 193–95

Foulke, William Parker, 9

France, 40–42

Germany, *13*, 99–100

Gilmore, Charles, 62–63, 65, 66

Gobi Desert, 46–47, 118–19

Golden Gate Highlands National Park, 48

Gorgosaurus, 55, 62–63, 65, 115–16

Gould, Stephen Jay, 141

growth changes. *See* ontogeny

Häberlein, Karl, 100

Hadrosaurus, 10, 31, 71

Hanksville-Burpee Quarry, 160

Hatcher, John Bell, 138

Hawkins, Benjamin Waterhouse, 9, 71

INDEX   223

Hell Creek Formation, 64, 67
Hesperornithoides, *104*
Hitchcock, Edward, 5
Hopkins, John Estaugh, 9
horns, 137–41, *137*, *140*, *142*, *146–47*, 148–49, 162, 199
hot-running dinosaurs. *See* physiology
Huxley, Thomas Henry, 101
Hylaeosaurus, 8–9
Hypselosaurus, 41–42

Ichthyosaurus, 6
Iguanodon, 8–9, *10*, *16*, 70–71, 74, 116, 120
India, 11, 47, 89, 95
Inostrancevia, 21
Ischigualasto Provincial Park, 25

"Jane," 64, *64*, 66, 67, 153
jaws. *See* teeth and jaws
Jeholornis, 129–30
Ji, Qiang, 103
Ji, Shu-an, 103
"Jingo," 167
"Joe," *56*, 57–58, 120, 148–49

Kammerer, Christian, 34
Kentrosaurus, 144
Kielan-Jaworowska, Zofia, 200–201
Kircher, Athanasius, 3
Knight, Charles R., 71, *72*, 75
Kongonaphon, 106–7
Kongonaphon kely, 34–35
Kosmoceratops, 140–41
K/Pg extinction, 171–79, *178*
Kwanasaurus, 26

Lambeosaurus, 56–57
Laramidia, *147*, 161–62
large size of dinosaurs, 85–96; armored dinosaurs, 144; evolution of, 87–93, *90*; fossilization chances due to, 184, 185; individual variation and upper limits, 96; "largest," 85–88, *88*, 94–96, *95*
Larson, Derek, 177
Laskar, Amzad, 81
Leallynasaura, xi
Leaping Laelaps (Knight), 71, *72*, 75
Lee, Andrew, 44
Lee, Yuong-Nam, 201
Leidy, Joseph, 9–10, 11–12

Lightning Ridge, 188
lines of arrested growth (LAGs), 44
Lockley, Martin, 153–54
Lyell, Charles, 166, 182

Madagascar, 34, 130
Maiasaura, *45*, 50, 122–23, 126–27, *126*
Mamenchisaurus, *90*, *95*
mammals: and air sacs, *90*, 91–92; definition and description, 134; diet and digestion by, 92–93; end-Cretaceous mass extinction survivors, 169, 174, *178*, 179; evolution of, 19, 20–23, 42, 134–36, *135*; misconceptions about, 131–35, *133*, *135*, 166–68, 169; parenting care, 89–90; reproduction, 89–90; size comparison to mammals, 88, 89–93, *90*
Mantell, Gideon, 8, *10*
Mantell, Mary Ann, 7–8
Maraapunisaurus, 95
Marsh, Othniel Charles, 11–12, 101, 137–38
Massospondylus, 48
Mbiresaurus, 29, 31
Medusaceratops, 140–41, 161
Megalosaurus, x, 7, *7*, 8–9, 70–71, 113, 115
Megatherium, 6
Mei long, 186
Mexico, 172
Microraptor, 105, *107*, 108, 111
Minmi, 123
Mongolia, 46
Montana, 62, 75–76, 122, 126–27, 138, 145–46, 151, 161, 174, 182
Morrison Formation habitats, 87, 93, 193–94
Mosasaurus, 6, 32
Mother's Day Quarry, 159
Muséum National d'Histoire Naturelle, 41
Mussaurus, 58
Mygatt-Moore dinosaur quarry, 117
Mymoorapelta, 144

Nanotyrannus lancensis, 63, *64*
Nanuqsaurus, 83
Natural History Museum, 25–26, 100
Nesbitt, Sterling, 27

nesting. *See* eggs and nests
Nigersaurus, *118*, 119
North Dakota, 174, 182
Nyasasaurus, xi, 107
Nyasasaurus parringtoni, 27–28, 31–32

On the Origin of Species by Means of Natural Selection (Darwin), 99, 101, 182
ontogeny, 53–67; ecosystem influences, 59–67, *59*; juvenile identification, 57–59, *61*, *64*; new species misidentifications due to, 53–59, *56*, *64–65*
ornamentation: armor, 143–49, *144*; feathers as, 108–11; future research directions, 199; horns, 137–40, *137*, *140*, *142*, *146*, 148; misconceptions about, 138–39, 140–41, *142*, 146–49, *147*
Ornitholestes, 14, 73–74, *74*
Ornithomimus, 108
Ostrom, John, 75–76, *77*, 156–58
Otero, Alejandro, 58
Oviraptor, 46
Owen, Richard, 8–9, 11, 29, 71, 99

Pachycephalosaurus, 31, 58, *146*
Pachyrhinosaurus, 83
paleontology: "Bone Wars," 11–12; and colonialism, 4, 10–13, *13*; Dinosaur Renaissance, 15–16, *77*, 78, 103, 155; evolutionary theory, 13–16; present-day and future research directions, 16–18, *16–17*, 84, 197–202; rise of, 2–9, 18; Second Jurassic Dinosaur Rush, 12
Pangaea, 20–23, 29, 35, 37
Parasaurolophus, 30, 55, *56*, 57–58, 82, 120, 148–49, 163
parental behavior, *41*, 46–47, 49–50, *50*, 89–90
Parrington, Francis Rex, 25–26, 27
Patagonia, 47, 89, 174
Patagotitan, x, 31, 81, 87, 95
Petrified Forest National Park, 25
Philpot, Elizabeth, 109
physiology, 69–84; digestion, 92–93; Dinosaur Renaissance and, *77*, 78; future research directions, 84; metabolism, 82–83; reptile stereotype, 69–73, *72*; reptile

stereotype exceptions, 73–74, *74*; revising assumptions about, 74–78, *77*; thermoregulation, 78–83, *80*, *82*, 91–92, 107–8, 143, 144
pigeons, 30–32, *30*
Plesiosaurus, 6
Pliny the Elder, 3
Pol, Diego, 58
Poland, 135–36
pollen, 123–24
Pouech, Jean-Jacques, 40–42
primates, 130, *178*, 179
Prince Creek Formation, 82–83
Procheneosaurus, 55–56
Protoceratops, 46
Psittacosaurus, *65*, 104, 108, *133*, 134, 201
pterosaurs, 24, 82, 106–7
Puertasaurus, 95
Purgatorius, *178*, 179
Pyrenees Mountains, 40

raptors, 25, 105, *107*, 108, 111, 156, 162, *175*, 176–77
Repenomamus, *133*, 134
reproductive behavior and strategies, 17, 39–51, 108, 153–55, *154*. *See also* eggs and nests
reptiles: evolution of, 19 (*See also* dinosaur family tree; *specific reptiles*); Permian period, 19–20; post-Permian mass extinction, 23; stereotypes, 69–73, *72*; Triassic period, 23–24
Roniewicz, Ewa, 200–201
Royal Belgian Institute of Natural Sciences, 74
Ruhuhu Basin, x–xi, 25–26, 27

Sanajeh, 49
Sarahsaurus, *29*
Saskatchewan, 115
Sattler, Bernhard Wilhelm, *13*
Schweitzer, Mary Higby, 43
Science, asteroid hypothesis article in, 169–70
Scutellosaurus, 143
Second Jurassic Dinosaur Rush, 12
Seeley, Harry Govier, 30–31
Seismosaurus, 95
Siberia, 21–22
Silesaurus, 27, 28

Sinocalliopteryx, *116*
Sinornithomimus, 160
Sinosauropteryx, xi, 103, 106, 108, 110
Sinraptor, 153
size. *See* large size of dinosaurs
social behavior, 151–63; courtship
   displays, 153–55, *154*; diversity of,
   162–63; evidence of, 152–54, *158*;
   feathers for, 108; herd and pack
   behavior, 75–76, 158–62, *158–60*;
   horns for, 139, 140, *142*, *147*,
   148–49, 162; of juveniles, 158–61,
   159–60; ornamentation and armor
   for, 143, *144*, 145–49; overview, 17;
   preconceived notions about, 151–52,
   151; trackway evidence of, 152,
   157, *158*
Society of Vertebrate Paleontology, 15,
   103, 201
South Africa, 11, 48
South Dakota, 174, 182
South Korea, 2, 155
Spinophorosaurus, *183*
Steensen, Niels, 3
Stegosaurus, 15–16, 30–31, 69–70, 86,
   120, 144, 167
Sternber, Charles H., 189
Struthiomimus, 200
Styracosaurus, 31, 32, 58, 138
Supersaurus, 87, *90*, 95

Tanzania, x–xi, *13*, 25–26, 27, 144
teeth and jaws: comparison to
   mammals, 92; and diets, 113–14,
   114, 115–19, 121, *126*; growth
   changes, 65; of *Megalosaurus*, 7, *7*
Tenontosaurus, 44, 156–58
Tetragonosaurus, 56
Texas, 159–60
thermoregulation, 78–83, *80*, *82*,
   91–92, 107–8, 143, 144
Torosaurus, *170*, 176
Torvosaurus, 94, 117–18, 194
Trachodon, 189
Transantarctic Mountains, xi
Triassic-Jurassic mass extinction,
   36–37, *36*
Triceratops, 12, 30–32, *30*, 74–75,
   104, *126*, 137–41, *137*, *140*, 160,
   170, 182
Triceratops horridus, 138–39

Tyrannosaurus, *17*, 31, 69–70, 94, 139,
   140, 193
Tyrannosaurus rex, x, 43–44, 62–67,
   64, *114*, 115–16, 141, 151–53, *151*,
   163, 182

United States, colonization of, 11–12.
   See also specific states
Utah, 2, 12, 47, 57, 85–86, *86*,
   117–18, 123, 133, 145, 174, 193

Vinther, Jakob, 109–10
von Meyer, Hermann, 100

Wallace, Alfred Russel, 99
warm-blooded dinosaurs. *See*
   physiology
Werning, Sarah, 44
Western Interior Seaway, 161, *165*,
   175–76
Whitlock, John, 60
Wiemann, Jasmina, 81–82
Wyoming, x, 12, 73–74, 138, 174

Yucatán Peninsula, 172
Yutyrannus, *80*, 108

Ziapelta, *144*
Zion National Park, 2
Zuul, 145–46, 148–49

About the Author

RILEY BLACK is the award-winning and bestselling author of *The Last Days of the Dinosaurs, When the Earth Was Green*, and many other books about fossils. She is a frequent contributor to publications such as *Smithsonian, National Geographic*, and *Slate*, and has made repeat appearances on programs such as *NOVA* and *Science Friday*. In 2024, she was awarded the Friend of Darwin award from the National Center for Science Education for her efforts in science communication. And when not enthusing over fossils for the public, Riley frequently joins museum field crews to discover new fossils across the American West.

rileyblack.net

Also available in the Shortest History series

Trade Paperback Originals • $16.95 US | $21.95 CAN

   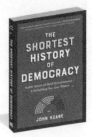

978-1-61519-569-5 978-1-61519-820-7 978-1-61519-814-6 978-1-61519-896-2

978-1-61519-930-3 978-1-61519-914-3 978-1-61519-948-8 978-1-61519-950-1

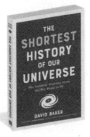  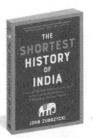 

978-1-61519-973-0 978-1-891011-34-4 978-1-61519-997-6 978-1-891011-45-0

978-1-891011-66-5 979-8-89303-060-0 979-8-89303-012-9 979-8-89303-052-5